B"H

Something wonderful
is going on...
and YOU can take an
important part in it.
What part?
That is up to YOU!
This Book Belongs to:

Activity Kit

for

The Milkweed, the Monarch, and the Moon

Your activity kit is also a coloring book.
Add colors, doodles, and drawings...

Learn and have fun!

ISBN-13: 978-0-9909000-3-0
ISBN-10: 0-9909000-3-7

Education/Elementary EDU010000

Activity Kit

for
The Milkweed, the Monarch, and the Moon

by
Davida Lippman

The creation of this Activity Kit to learn more about the ideas in The Monarch, the Milkweed and the Moon book took teamwork, so...

Special thanks to all the true friends who gave genuine, generous support to make this project a success.

Great appreciation to Dee Teller for her enthusiastic attention to make this book more fun for you!

Rebecca Zwiren used her professional skills to design the book cover and to prepare the pages to be published.

Most of all, YOU make all we did worth our effort, as you learn from this book, so the MMM team sends you a cheerful

Is there someone that you would like to thank for helping you? This page was designed to write your own Thank You note... clip it out and deliver it.

Hi, I'm Davida

Hi, I'm Dee

Do you dream of doing things that seem impossible?

This activity kit was created to encourage you to use the spark of your imagination for bright new ideas.

You will practice ways to sort information, to make choices and to follow a plan to make the world a better place.

The best part of learning is to discover more about who you really are, because that makes everything you do a valuable experience!

Ready to begin?

Jump in!

Contents | Page

The time has come to launch your learning adventure!

Enjoy progress at your own speed!

3... 2... 1...

The Beginning of Section 1

Check out the cover of the book "The Milkweed, the Monarch and the Moon" **Let's call it "MMM"**.

Mark an X beside the description below of what you expect the book to contain.

[ ] information about nature (Science - Nonfiction)

[ ] a make-believe adventure in a fantasy world (Science - Fiction)

[ ] both non-fiction and fiction

The cover has pictures of real natural things, so you can expect the story to give information (Science - Non-fiction.) However, some fantasy (Science - Fiction) was added just for fun, so...

The correct answer is: both non-fiction and fiction!

For example, look at the cover and answer these questions-
Does the moon *really* smile? ______
Do caterpillars *really* reach out to hug their parents? ______

If you answered "no", then great! you're able to tell the difference between what is real and what is fiction. That is called being able to "discern." Someone who cannot tell what is real or fiction is called "naïve" (niy-yeev.) It is important to know what is real...and, it is fun to play make-believe!

Discern... discern... discern...

Name a story you read that is fiction and one that is non-fiction.

Here are more words to describe either kind of story.
(1) **fiction** = pretend, imaginary, fake, whimsical, make-believe.
(2) **non-fiction** = accurate, factual, authentic, historic, scientific.

Write the name of a fiction (F) book you have read below.

____

Next, write the title of a non-fiction (NF) book you have read.

____

Write F or NF in each box below to show you discern the difference.

[ ] The monarch butterfly has a lifecycle that starts with an egg.
[ ] Squirrels use tools to take care of a garden.
[ ] The monarch caterpillar eats the milkweed to grow.
[ ] Snails and turtles make plans to go on playdates.

Color… color… color…

Book Cover Activities

1. Fill in the blank labels

The copy of "The Milkweed, the Monarch and the Moon" front cover on the opposite page has blank labels and missing colors. Make the page complete by following the directions below.

- Use the blank labels to write "plant" or "animal." Some labels apply to two of the same kinds of things. For example, the mushrooms and grass are both plants.
- Add color. The flower petals may have any color, but the stem and leaves should be green.
- The moon is already labeled "Inanimate Object." It is not a plant or an animal. Try to remember this information.

2. Match words and definitions

On the back cover is a blurb. This gives information about what's inside. There are three words in the blurb that are important to understand: **observe**, **track**, and **spend time**.

Draw a line to connect each word to the definition that is a match.

observe	to check on progress
track	to watch something happen
spend time	to be involved in an activity

Clue: observe is another word for to watch

Think... think... think...

Don't be like a robot....  **Think!** 

Long ago, before electricity was invented, there were no computers to help get information. People could only watch and listen to what was naturally happening. If that is hard to imagine, maybe take a camping trip with your family.

Try to have a more positive attitude (courageous, forgiving, hopeful...) Try to be less negative (afraid, ashamed, doubtful, angry.) Practice thinking "good stuff" and you may come up with an idea to improve things!

Quote from a famous inventor

"Thinking is the hardest work there is, which is probably why so few engage in it."

- Henry Ford

Word Puzzle

Unscramble the letters - E S R B V E O _ _ _ _ _ _ _

Clues: This word is in the blurb. It starts with an O.
Try to figure out this word before you find it in the sentence below.

Circle the photo of a natural phenomenon below that you think would be the most fantastic event to observe.

A rainbow appearing

An egg hatching

A bolt of lightning

An icicle forming

A flower blooming

A spider making a web

Messages from the Author

Dear Reader,

Please see the messages that I wrote inside the cover. Thanks!

Faithfully, Davida

My first message - The copyright notice

© is a symbol that shows permission must be given to make copies. I allow anyone who buys this book to make copies of my work for the purpose of education. Thanks for buying my book!

Use the line below to name a thing that you made or that you own. It is your property.

____

Who would you allow to use this property?

____

What property did someone else own and give you permission to use?

____

Try to remember.... It is right to take care of someone else's property the same way you take care of what you own!

My second message - The dedication

This book honors my mother, Carmela, because she taught me to find refreshment in the beauty of nature.

In honor of Carmela
Someone very special

Write the name of someone that you would like to honor on the line below.

____

This person is special to me, because . . . (write a sentence below)

____

____

____

In the frame below, describe, draw, or paste a photo of the person you chose.

MY CHOICE

RESPECT

APPRECIATION

HONOR

Take a moment to consider the list of honorable personal qualities below. Circle at least **five** words that describe you.

You may draw or put a photo of yourself here:

patient

kind

generous

careful

honest

talented

funny

neat

hardworking

unselfish

forgiving

polite

Another thing I like about myself is that I am -

Would you like to take a break and stretch? You are about one third of the way through the first section...

Consider... consider... consider...

Nobody is perfect!

Keep your courage and learn from mistakes!

Remember... remember... remember...

Use what you already know about plants and animals to circle T for True or F for False after you read the statements below.

Plants have roots.	T / F
Plants have a brain.	T / F
Animals and plants have life cycles.	T / F
Animals can make food from sunlight.	T / F

Plants and animals are living things. That makes them the same in some ways. For example, they both have life cycles. The rest of the things you find around you are inanimate objects.

Is a shell a living thing? No . . . Although an animal grew the shell, it was never alive itself.

Make a list of five items around you that are inanimate objects:

A lightbulb

1. ____________________

2. ____________________

3. ____________________

4. ____________________

5. ____________________

Cycles... cycles... cycles...

The MMM book tells that a monarch's life begins as an egg. Next, a caterpillar hatches and grows. After that, a metamorphosis (complete change) takes place inside a chrysalis and out comes a butterfly!

Can a caterpillar return to an egg or can a butterfly go back to the form of a caterpillar? ...of course not! The life cycle of the monarch butterfly is a pattern that repeats in the next generation.

Use the word bank to label the pictures of the monarch life cycle. The pictures are in the correct order, but the words are not!

Word bank: egg, chrysalis, butterfly, caterpillar

Draw a line to connect forms that are part of a species' life cycle:

puppy	chicken
chick	frog
baby	dog
tadpole	child
lamb	cat
kitten	sheep

Wonder... wonder... wonder...

Is there an activity that you do just for fun? This is called having a hobby... For example, you may find wonderful inanimate objects like shells or rocks around that are safe and free to collect. If you have a hobby, or have an idea for one, tell about it below.

____

____

____

____

____

Mystery Word

The code:

 = S

 = T

 = R

 = O

Use the code to make a word below the pictures

 +  +  + 

_____ _____ _____ _____

The answer is in the sentence below:
Scientists use definitions to sort different kinds of things into sets.

A Sets Game: Living Things

On the next page there are pictures of roses, reptiles, dogs, ducks, and humans. Each kind of thing has three pictures that make a set.

1. Cut out the cards and shuffle them.

2. Place them face down in three neat rows of five cards.

3. Pick up any two cards. If they belong to the same set, keep them face up in the same place, until you find the third one.

4. If they don't belong to the same set, turn them over again and try to remember where they are.

5. Once you find the three cards for a complete set, pick them up.

6. Continue until you have picked up all five sets.

By the way, the ducks in the pictures were my pets. When their eggs hatched, they moved to a farm near a lake, because my yard was too small for their growing family. I hope they like their new home!

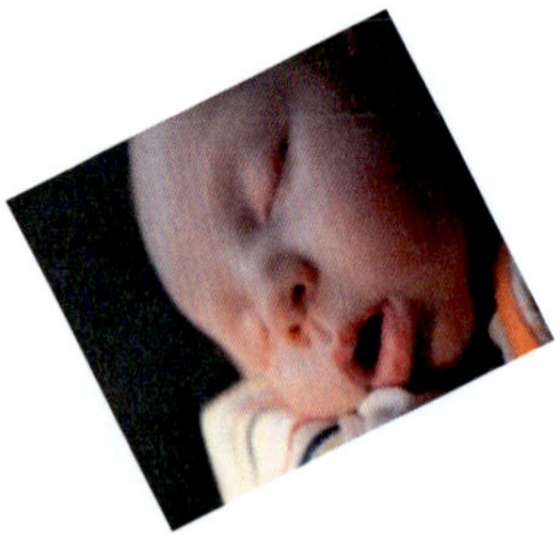

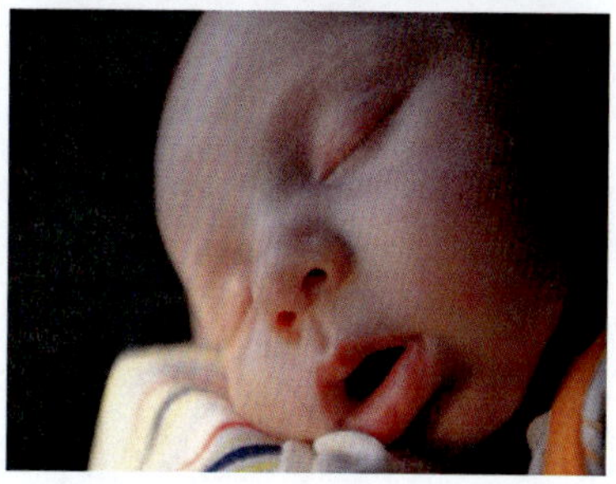

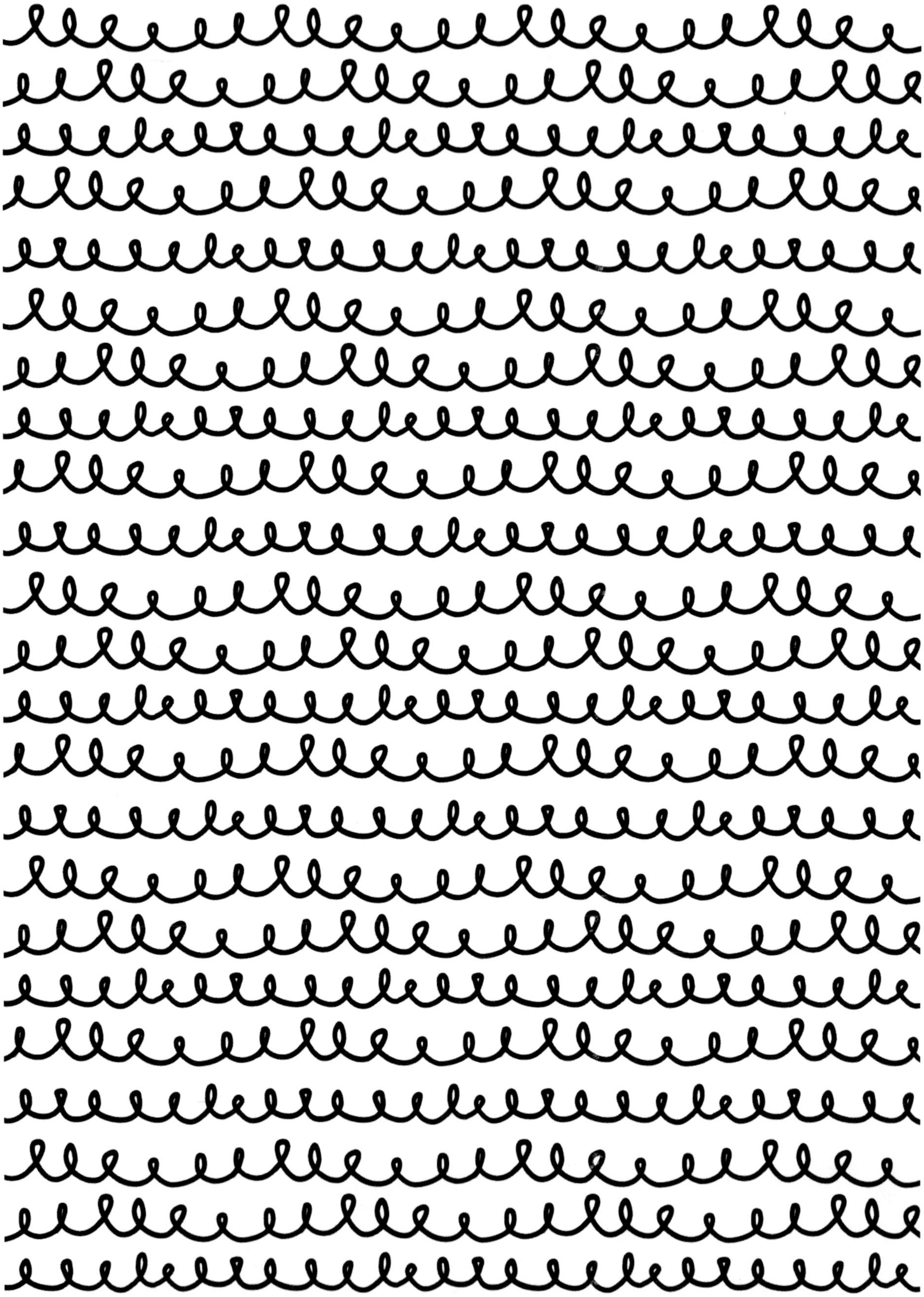

Sort... sort... sort...

Use the three definitions on this page to sort the kinds of things on the next page. Each kind of thing has four pictures. Put the name below it in the correct column of the chart on the next page.

Plant

This kind of thing must eat. It can make food from sunlight. It grows and changes in a lifecycle. New plants can come from seeds, roots, or stems. It has a kind of intelligence, so it can sense changes, but it doesn't think, because it has no processor, like a brain.

Animal

This kind of thing must eat, but it cannot make food from sunlight. It grows and changes in a lifecycle. New animals come from eggs or live births. It has intelligence and can think with a processor, like a brain. These can learn from experience.

Inanimate Object

This kind of thing never was alive. It may require fuel or a battery. However, that is not like eating. It does not grow and change. Some of these things can be programmed in a central processing unit, like a robot, but that is not like having its own brain to think and learn.

If you like doing math, fill in the answer to this question... How many pictures are there? (3 kinds x 4 pictures of each kind = ____)

Chart of Three Kinds of Things

Plant	Animal	Inanimate Object
Rose	Horse	Moon

Pineapple

Butterfly

Bicycle

Rose

Cactus

Airplane

Fish

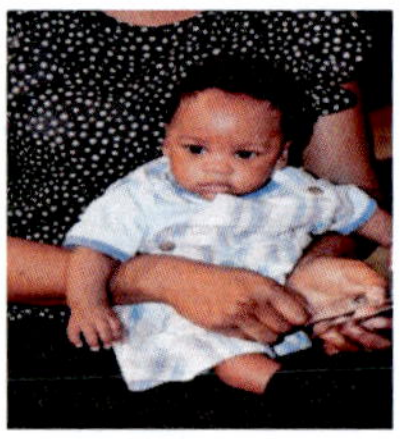

Human

Grapevine

Horse

Shoe

Moon

Sort... sort... sort...

You already sorted plants, animals, and inanimate objects by a definition of what is different for each one. What if something is in common? For example, both plants and animals need water.

To show what is common between two or three kinds of things, read each statement on the chart below and put an X where it fits. Use **P** for Plant, **A** for Animal, and **IO** for Inanimate Object. For example, Statement #2 fits P for Plant and A for Animal. Do your best, then check your answers on the next page.

Statement Chart: Kinds of Things

#	Statement	P	A	IO
1	This thing can make food from sunlight			
2	These need water to survive	X	X	
3	This never was alive			
4	New ones come by eggs or live births			
5	New ones come by seeds, roots, or stems	X		
6	New ones do not come naturally			
7	These exist in the natural world (non-fiction)			
8	This gathers information and thinks			
9	This does not have a brain			
10	This grows and changes in a lifecycle			
11	This could possibly be trained, if it is alive, or could be programmed to do something, if it never lived			

Answers - Statement Chart: Kinds of Things

#	Statement	P	A	IO
1	This thing can make food from sunlight	x		
2	These need food and water to survive	x	x	
3	This never was alive			x
4	New ones come by eggs or live births		x	
5	New ones come by seeds, roots, or stems	x		
6	New ones do not come naturally			x
7	These exist in the natural world (non-fiction)	x	x	x
8	This gathers information and thinks		x	
9	This does not have a brain	x		x
10	This grows and changes in a lifecycle	x	x	
11	This could possibly be trained, if it is alive, or could be programmed to do something, if it never lived		x	x

Directions are given on the next page to make a Venn Diagram

Venn Diagram: Kinds of Things

A Venn diagram shows a circle for each thing. What is in common gets a mark in the area where the circles overlap.

To make a Venn Diagram, follow these directions -
Refer to the completed statement chart. Take the number of each statement on the chart and put it on the diagram where it belongs.

Does it go in one circle or where the circles overlap? Use each number once to fill all the blank circles.

Two examples are given. Do your best before you turn the page and see the answers.

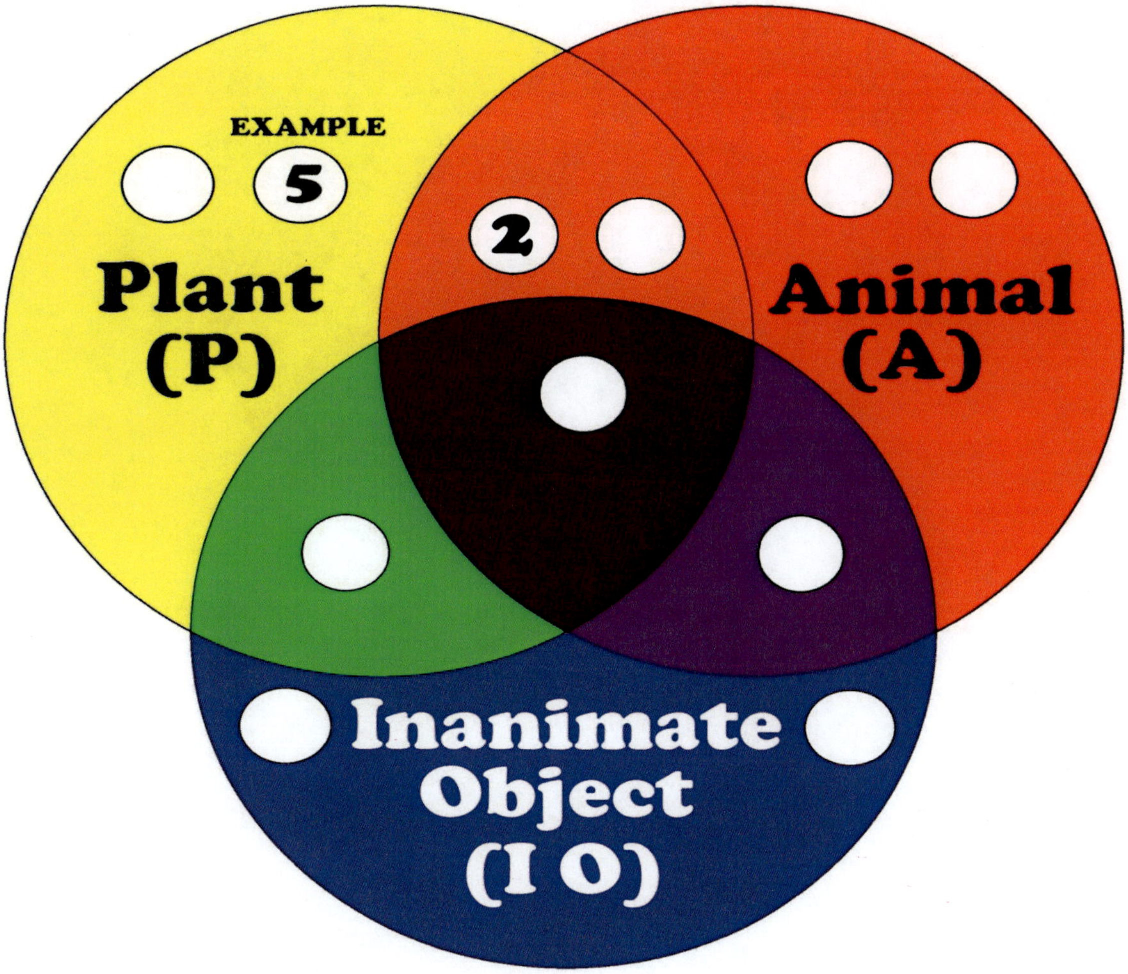

Answers - Venn Diagram: Kinds of Things

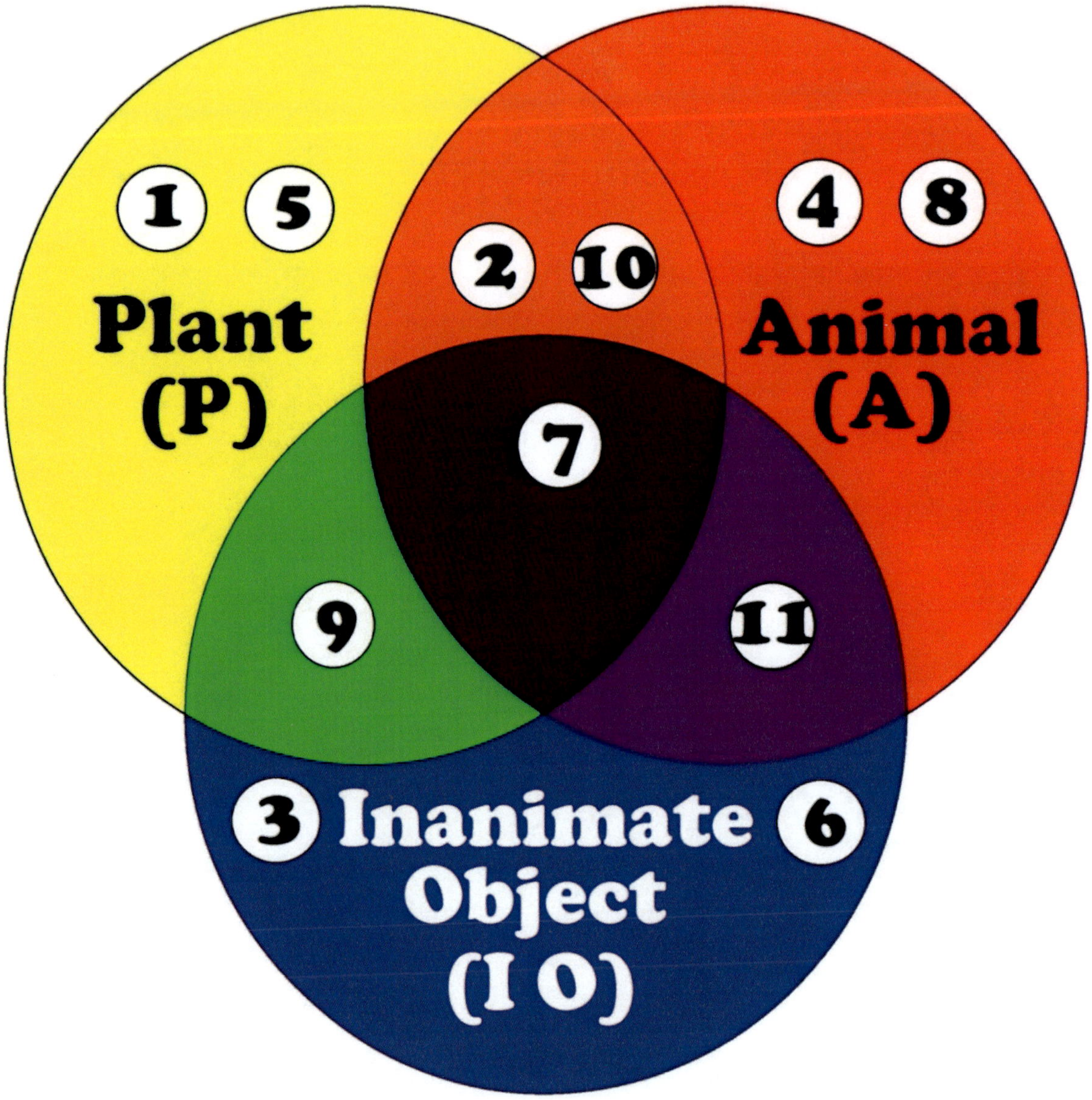

Consider... consider... consider...

Do you already use machines at home to do laundry or take care of other routine chores? Are there things you have to do, that you'd rather tell a robot to do?

Here are some reasons to give a job to a robot-
A robot can not complain. It doesn't worry. It can give the same results every time without feeling bored.

Here are some reasons not to give a job to a robot-
A robot is limited to what it can be programmed to do. It can need unexpected repairs. It doesn't care about teamwork.

Read the following scenerios to consider when you would give a job to a machine. Think freely. Every opinion deserves respect!

(A) In a family with tired parents and children who didn't learn to read yet, would a robot to read a bedtime story be nice?

(B) When a job is dangerous, would it be better to program a robot to do it, so there is no risk of harm to a person?

(C) Scanners can replace cashiers and make lines short at the supermarket, but a neighbor will lose his job if people use them. Should you wait in line or use the scanner?

Imagine... imagine... imagine...

What if you had to decide if you would own a garden, a bicycle, or a puppy. First, you can gather information about these things. Then you can sort out what makes them different, to know which one suits you the best.

Next, read the pros (something good) and cons (something not so good) about owning these three different things. Lastly, make your decision. Your choices are part of what makes you unique. Enjoy yourself!

Garden
Pro - A garden can add beauty to your home.
Con - You must give plants water and pull weeds.

Bicycle
Pro - You can leave a bicycle alone when you don't use it.
Con - You must follow safety rules and take care about repairs.

Puppy
Pro - A pet can show you appreciation for your care.
Con - You must feed it and clean up after it every day.

On the next page, try to explain how you feel about your decision.
Here are some verbs (action words) for making your choice:

pick, select, prefer, favor, opt for

Here are some nouns (name words) to call the thing you choose:

choice, selection, preference, option

Pros

Cons

Did pros and cons help you to think about which kind of thing you would rather have? Express your thoughts about your choice below. Use a noun and a verb from the list. For an example - My preference would be to own a ____. I opt for a pet because _______.

____

____

____.

____

Can you think of another pro and con for your choice?

Another pro is ____________________
Another con is ____________________

You will learn from every choice you make. It is part of growing!

The value of anything depends on how much you care!

In 1922 Marjorie Williams published a story about a boy who cared so much about his toy that it became real.

New readers still enjoy the way this book shows the value of caring -- even after 100 years, which makes it a "classic" story.

Maybe you have a real friend that you care about...

A quote from a famous boxer:

"Friendship is the hardest thing in the world to explain. It's not something you learn in school. But if you haven't learned the meaning of friendship, you really haven't learned anything."

- Muhammad Ali

A dog's brain is at the level of his heart, but a person walks with his head higher. Let your brain choose who has a place in your heart!

Five ways your brain gathers information:

Write a sentence on the lines below for each sense in the picture on this page. The word banks contain all you need to make five matched pairs. Use each word one time. Then name something that you like to use this sense for. An example is given for you to use as a model.

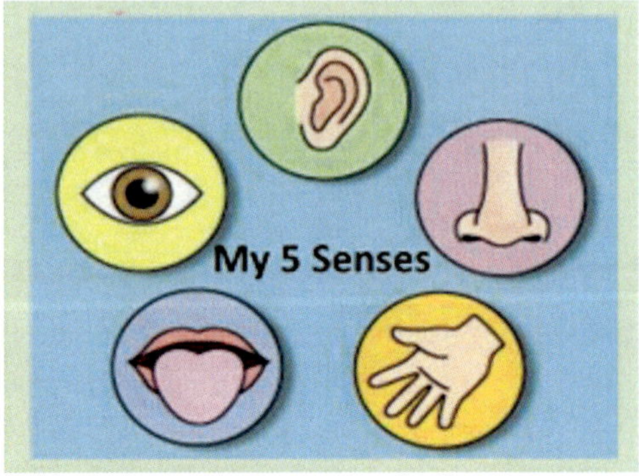

My nose gathers information so that I can smell things. For example, I like to smell a rose.

Word Bank 1 - What to use to gather information: eye, ear, nose, tongue, skin (especially fingertips)

Word Bank 2: hear, see, feel, smell, taste

1) ____

____

2) ____

____

3) ____

____

4) ____

____

5) ____

____

Just for fun, a riddle: What word found in the word banks has - 3 letters, - starts with an "e" and ends with an "e" - and when you hear it, this word sounds like you're saying a letter of the alphabet?

— — —

Think of what senses you would use to know more about the things in these pictures.

Challenges... challenges... challenges...

Sometimes a limited ability, like blindness or deafness, presents a special challenge. That limit may be called a disability, because in that case, an extra effort is needed to succeed.

A few of the famous people who faced special challenges with great success are pictured below. Their examples of endurance give us all encouragement to never give up!

Hellen Keller
(1880-1968)

Stephen Hawking
(1942-2018)

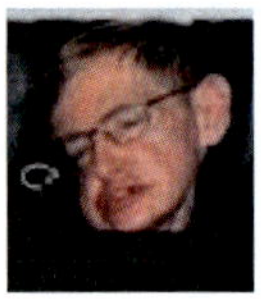

Chris Gardner
(1954- alive)

Helen Keller was blind and deaf. She showed everyone not to expect less from someone who has a special challenge. One of the movies about her life is called, "The Miracle Worker."

Stephen Hawking was disabled and used a wheelchair, but he continued his work and made scientific discoveries that changed how we think about our universe. "The Theory of Everything" is a movie about him.

Chris Gardner had a lack of opportunities because of poverty. By his hopeful perseverance he achieved great wealth and helped others. This inspired a movie called "The Pursuit of Happyness."
(Please note, the word happiness is intentionally spelled incorrectly)

Write about how you faced a challenge, for example to become able to swim. Use the words "self-confidence," "courage," and "perseverance" to explain how you continued, even if you almost gave up. Try to remember how you kept your hope for success through difficult times. After you write about a good experience, you can read your story for inspiration the next time you face a challenge!

How I Faced a Challenge

by (You are the author! Put your name here.)

Wonder... wonder... wonder...

All animals use the information that their senses tell them.

Some animals have more wonderful senses than humans...

- A bat can see in the dark using echo sound waves
- A whale can send signals through deep water
- A bear can smell food from great distances

What makes you so special is how you get creative ideas.

Only humans have the creativity to change our world, for example to build bridges!

All around the globe people share ideas to take care of our world.

Scientists and farmers work together to find the best way to produce food for the world's growing population.

You are about two thirds of the way through the first section.

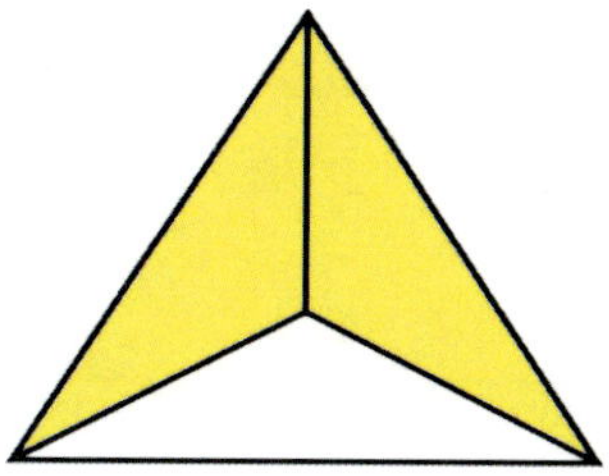

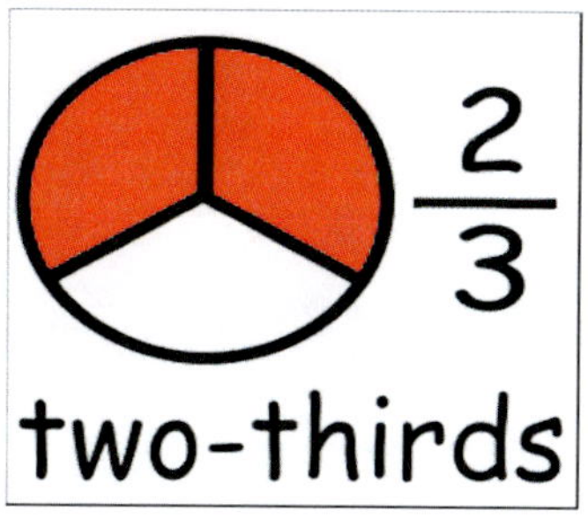

Team Work

When different kinds of people work together on a team they have more success! This is called cooperation.

Here are four basic kinds of job positions:

The creative thinker creates something new

The leader leads everyone to accept the new idea

The organizer organizes everything for the work to be done

The producer does the work to produce the results

Longer descriptions of the job positions:

The creative thinker uses imagination to make an invention or create art, like songs and books. Another word for creative is innovative. A pro to having this position is coming up with new ideas. A con is the idea may not be accepted by others.

The leader encourages others to accept a new idea. To succeed in this position a person needs to be trusted so others will follow. A pro is that leaders have the authority to use some power. A con is that not everyone will listen to the leader.

The organizer makes the plan. People in this position plan how to do things and then give supervision so things go smoothly. Organizers watch details. A pro is fixing problems. A con is that others want the organizer to fix everything quickly.

The producer is trained to do a specific job. People in this position include farmers, chefs, sports stars, and factory workers. A pro is knowing exactly what is expected. A con is stress for meeting the expectations on time.

Which description sounds the most like you? Remember -- you can grow and change. You may even have a metamorphosis!

Scenarios... scenarios... scenarios...

The next few pages describe three fictional situations, also called scenarios. These scenarios must use teamwork to fix a huge problem. Fill in the blanks with the position that would do that job.

Remember:

The creative thinker creates something new.

The leader leads everyone to listen about what is new.

The organizer organizes what needs to be done.

The producer does the job to make the product.

Scenario 1

An earthquake destroyed a city's electrical system. The community depends on electricity to stay warm, to cook, and to use phones. They need a new electrical system!

The____________ in this case, an electrical engineer, invents a new system for everyone to connect to a source of electricity.

The____________ needs to organize the facility, supplies, and the way to train the work force.

The ____________ needs to lead the community to gain support and build confidence that everything will be okay.

All the ____________s will learn the skills to produce and install the new system.

Clue: In the first scenario, the first step requires creativity!

Scenario 2

A long, dark winter has made a whole village grumpy. The families need a way to feel happy and to have fun again.

The _____________ makes a public announcement for singers and musicians to follow directions and enter a happy song contest.

The_____________ with musical talent must work together to create the happy songs.

The_____________ of the competition organizes the rules for the contest to run smoothly.

The_____________, in this case a music studio, will produce and distribute the winning songs, so the people can listen and sing along.

Clue: In this scenario, leadership must take the first step.

Scenario 3

A contagious disease made animals and people feel sick. Medicine is needed to make them feel well again.

The________________ , in this case laboratory researchers, conduct experiments and discover a cure.

The________________ announces the news to lead investors and managers to support the discovery.

The__________________ organizes everything to make the factory ready.

The_________________ will produce the medicine so the people can be cured.

Clue: The first step in this scenario took creative thinking.

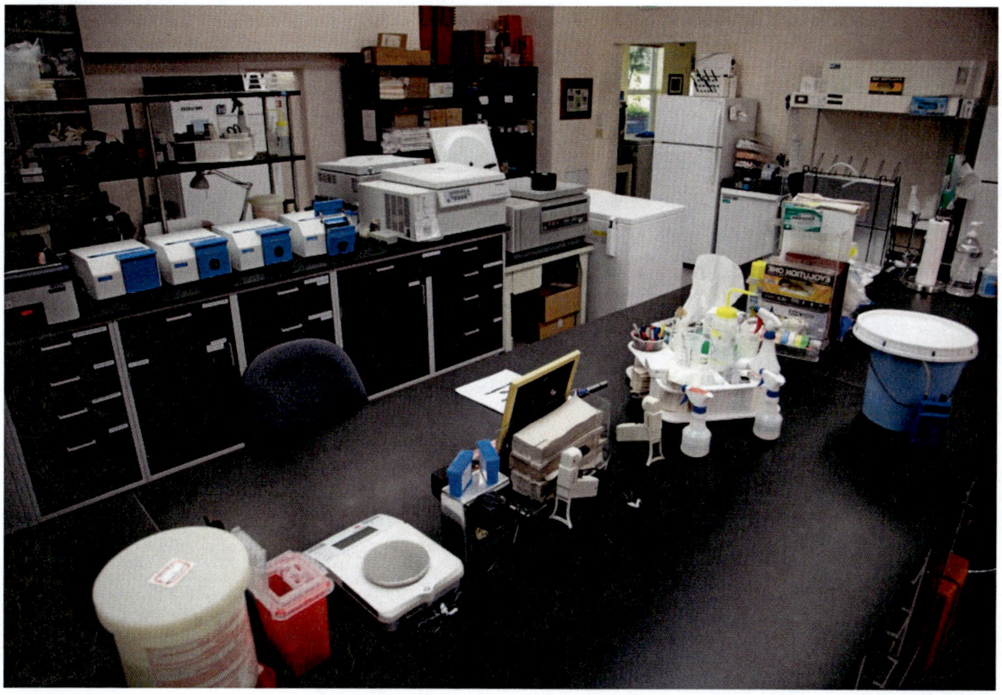

More Choices... more choices... more choices...

Complete the statement below. It may help to think about the work that you saw someone doing, that makes you wonder if you want to be in that kind of job position.

One job that I would like to try is _______________.

Write about what would you like to get done in that position and how you would work with other team members.

____

____

____

____

____

____

____

Mystery Word

Use this Letter Code to decipher an answer the question below:

What is crucially important for success?

Being part of a __ __ __ __

In the 1st section we studied:

- information about **The MMM** book
- life cycles
- ways you can gather and sort information
- kinds of job positions that work together

The next few pages have word puzzles to use what you have learned.

Have fun!

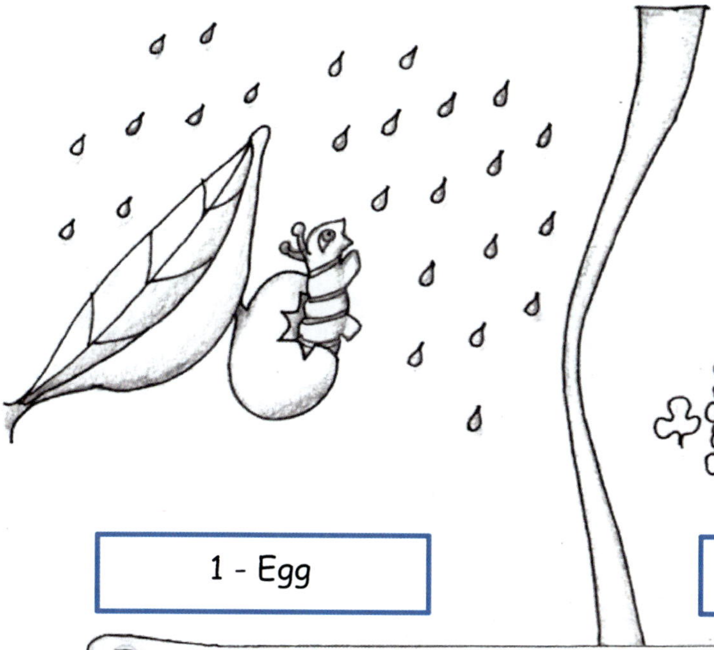

1 - Egg

2 - Caterpillar

Take a minute to color this page about the 4 life stages of the monarch butterfly

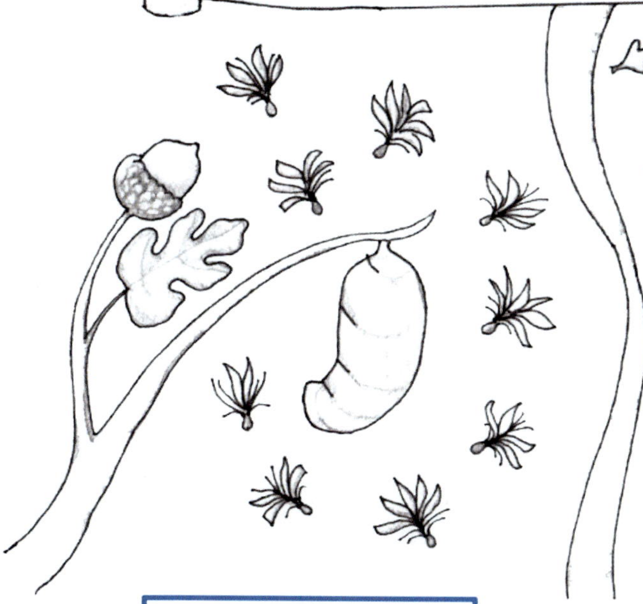

3 - Chrysalis

4 - Butterfly

Crossword Puzzle

Across

3. a type of colorful butterfly that begins its life on a leaf and is toxic to other insects
5. the natural object that revolves around the earth in a pattern that helps to observe that time is being spent
6. parts of a plant designed to catch sunshine for food production
9. part of a plant that connects from the roots to the other parts
10. a pattern that begins a new life over again
13. living thing that makes its food from sunlight
14. the first stage of a monarch butterfly's lifecycle
16. the stage of a lifecycle that is capable of reproduction
17. the color of the plant that makes food from sun rays

Down

1. living thing that gathers information and then processes it
2. the Earth and everything that lives on it (hint: starts with a "W")
3. a plant that is required for a specific butterfly to begin its life
4. a decision made by having a preference
7. based on facts that can be proven over and over again
8. a story that is not only based on facts, rather parts are imagined
11. the authority to give permission to make copies of a book
12. a few words on the back of a book to tell about what's inside
15. an ability to gather information, most humans can use five

See how well you can do before you look at the answers on the next page.

Check Your Answers:

Here's something you may also find interesting:

A man named Fibonacci found the same mathematical pattern in a conch seashell, a wave, a pineapple, and other natural things. He suggests that the pattern is the signature of who created these things!

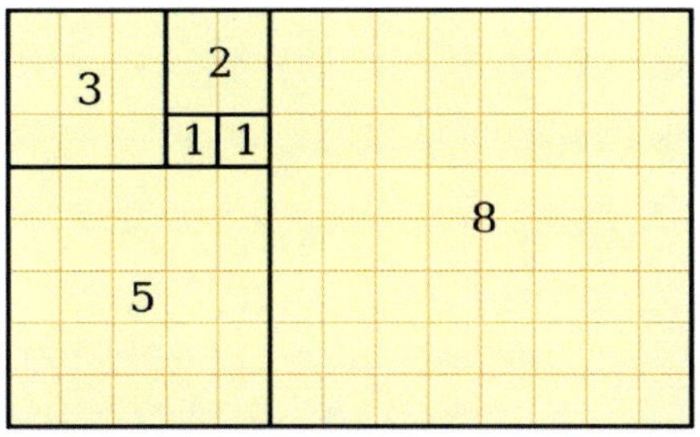

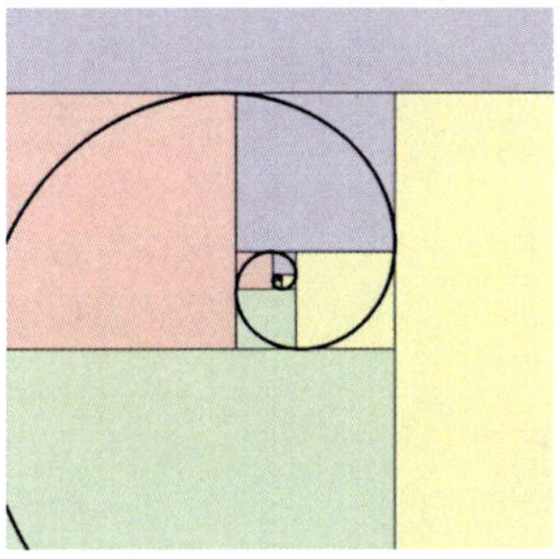

You might like to search for patterns in nature, like the placement of seeds in a piece of fruit or the colors on a bird's feathers . . .

You have reached the end of Section 1!

The Beginning of Section 2

Now that you learned something about doing things in Section 1, let's look at where on earth you can do them. Another word for a place to live is a **habitat**. Where you are effects how you live.

The monarch's habitat in "The Milkweed, the Monarch, and the Moon" (MMM) has a cold season. That is why migration is part of the story on page 14 of the MMM book.

The monarchs that live in Hawaii have no cold season, so they don't migrate. Even if you control the temperature inside your home, the natural climate still makes a difference in the world around you.

Put an X next to the description that is most like where you live.
[ ] In my neighborhood the trees get new leaves in the spring and in the fall they change colors. (deciduous forest)
[ ] There's hardly ever rain in my area. The temperature can be hot in the day and then becomes much cooler at night. (desert)
[ ] Not many plants grow in my cold and dry climate. (tundra)
[ ] Herds of cattle graze in my area. There's enough rain for many plants to grow, but there are not many trees. (grassland)

Do you wear different clothes in different seasons? Explain below.

____

____

FOREST DESERT

GRASSLAND TUNDRA

WHERE ON EARTH?

Every habitat has challenges to sustain a healthy environment. The seven largest areas on Earth are called continents. These are named **Africa, Antarctica, Asia, Australia, Europe, North America, and South America.** These are all on the map below.

My team members and I, Davida, all live on the same continent, in the same country and in the same city. I drew an arrow to it. Do your best to aim an arrow to the place where you live.

We live here. (Add your own arrow) **I live here.**

Tell... tell... tell...

Please write at least one complete sentence to tell about something you can do outdoors that gives clues about your habitat.

The point is to think of an activity that belongs to your natural habitat. For example, can you ski, go fishing, or climb a mountain?

____

____

____

Unscramble these letters for a word to describe an animal you can see without going to a zoo. A I N T E V = _ _ _ _ _ _
Did you get it? The word is in the sentence below.

One animal that is a native in my habitat is the peacock.

Tell... tell... tell...

Complete the sentences about things you can observe in the area around your home. You may use the words in brackets or your own words.

One animal that is native in my region is a ________________ [giraffe / pigeon / squirrel / seagull / monkey/ fox / cat / scorpion / rabbit /zebra] but I'd have to visit a zoo to see a ____________.

In my neighborhood, many people wear __________, but hardly anyone needs to own a pair of _______________.[snow boots, /beach sandals, /open sandals, /mountain hiking shoes/ closed shoes]

In less than an hour, I can take a ride to have a view of a _______, but I would have to travel much further to see a ________.
[a beach, /a mountain / a waterfall / an ocean / a river / a cornfield / a rice field / a sand dune / a forest/ a glacier/ a desert / a lake / a city]

Color... color... color...

Use orange and yellow to color the black and white picture. Page 13 of the MMM book explains why the colors of a monarch are bright.

The following story is based on the facts found by an international team of scientists. They discovered that an incredible navigation system is used for the butterfly to migrate. Some fiction was used to make the insect personal with a name and family relationships.

Zippy Goes with the Flow

One morning in Canada a majestic butterfly named Zippy came out of her chrysalis. Her special antennae picked up a warning signal that the climate is about to change. She will need to eat flower nectar for fuel. Going with the flow of a warm breeze will help Zippy to save her energy.

Zippy's journey will take two months to travel from from Canada to Mexico. That is 3000 miles!!! A monarch that is born at the end of the summer lives longer than usual. She will not be an adult to lay eggs until after she migrates. Then she'll retire in that warm climate. Her children and her grandchildren will fly north.

Even though Zippy's children were not born in Canada, her heritage will continue. Her descendants will return to the North and begin the cycle again!

Now use the lines below to tell something you found interesting.

____

____

____

____

Think... think... think...

Read the following paragraph and pay attention to the information about the numbers. Then answer the question below.

One milkweed had eight leaves. A monarch butterfly laid one single egg on each of the eight leaves (1 leaf x 1 egg x 8).

A single caterpillar crawled out of each one of the eight eggs. However, two didn't find a place to make chrysalises. (8 minus 2).

A storm came and blew away half of the chrysalises before they had a metamorphosis. (6 divided by 2)

How many butterflies were left?

Put an X by the correct number:

[ ] 8 [ ] 12
[ ] 3 [ ] 4

Sustainment is crucial for survival.

Air, water, nutritious food, and a home that gives shelter against any harm is needed to sustain life. Humans also need sleep, cleanliness, personal contact and some alone time to stay well.

Keeping these things in order is called sustainment. Some problems with sustainment are waste, pollution, and over-use of resources. Here are three words that start with R to help fix these problems:

Reduce - Use less of a resource. For example, turn off lights and other electrical things when you do not need them.

Reuse - Some things are still useful when they have finished their first purpose. For example, wash a container to refill it again.

Recycle - Many materials can become new things instead of waste. For example, newspapers can be processed to produce other pages.

Add your ideas to the sustainment chart below:

Recycle	Reuse	Reduce
Collect soda cans	Wash plastic containers	Use garden sprinklers

Environments, Ecosystems, and Biomes

Our planet has areas with different conditions that determine which plants and animals naturally live there. These are called the natives to their environment. Before we explore this, learn the words below.

Environment - all the features of a place, including the climate, the amount of water, and the landscape, such as mountains or rivers

Ecosystem - how all the native plants and animals of an environment connect with each other during their lifecycles

Biome - the way an environment and the ecosystem work together to sustain life

Environment + Ecosystem = Biome

Remember... remember... remember...

Use what you already know to put a **T** for True or an **F** for false after the following 5 statements. The answers are on the bottom line of this page.

1. Humans can make a home in an aquatic biome. [ ]
2. The ecosystem of a desert includes cactuses and lizards. [ ]
3. Animals in the tundra need insulation like a thick fur coat. [ ]
4. Tropical birds like flamingos are natives to the grasslands. [ ]
5. Tall trees and thick vines are features of a rainforest. [ ]

An aquatic biome, which includes lakes, rivers, coral reefs, and oceans, is not suitable for humans. Some people build their homes on stilts in aquatic areas, but this is not actually living in the water.

Answers to questions: 1 - F, 2 - T, 3 - T, 4 - F, 5 - T

Aquatic animals can breathe in water, but humans drown if they cannot breathe air!

Describe some features of the place where you live. Are there forests or are you near the coast of a salty ocean?

____

____

____

____

Here are five examples of different biomes:

1) **Aquatic** - Water covers this area. This includes lakes and oceans. Things that live there can take their oxygen from water.

2) **Forest** - Trees cover most of the area along with many kinds of plants and animals. This is the most common type of natural habitat.

3) **Desert** - Hardly anything can survive in the extreme dry heat of the day or grow in the sandy soil.

4) **Tundra** - These are very cold places, but a few plants and animals can live there.

5) **Grasslands** - This area has only enough fertile soil and rain for grass to grow, but not enough for trees. This is good farmland.

Use the descriptions on the previous page to label the 5 biomes below. One letter is given for a clue.

t _ _ _ _ _ _

_ q _ _ _ _ _

f _ _ _ _ _

d _ _ _ _ _

_ _ _ s _ _ _ _ _ _

Explore... explore... explore...

What a wonderful world!

There are more than seven billion eight hundred million human beings alive today... **7,800,000,000**! That's more than have ever died. Every single person is a part of your extended community.

The first team we belong to is our own family. After that we become aware that each of us is a member of a neighborhood, and a citizen of a nation, and from there we connect with the rest of the world.

Families around the world have different holidays and traditions. History helps to explain why there are various cultures, languages, and beliefs, based on what has happened in the past.

This book focuses on how you can make things happen, so lessons about what already happened belong in history books. To consider what you may do, let's think about the effect of natural habitats.

Explore... explore... explore... (continued)

Here are some imaginary friends to help you pretend to explore different biomes. They have names that are common in their native lands. Read what they have to say about their homes.

Alasie is a girl whose name means honest. She lives in Inuvik, Canada. Her biome is tundra. Alasie says, "My garden grows in containers because there's no fertile soil. We must watch out for hungry bears."

Batbayar is a boy whose name means joy. He lives in Darkhan, Mongolia. The area he lives in is a grasslands biome. Batbayar says, "Deer graze in grassy fields near my home."

Rafael lives in Manaus, Brazil, in a rainforest biome. Rafael means "God heals" His kind of forest has tropical trees and vines. Rafael says, "My garden stays wet with dew and frequent showers. There are wild monkeys and colorful parrots in the trees."

Zidan is a boy whose name means "Delighted." He lives in Tamanrasset, Algeria. That is an area whose biome is a desert. "My garden mostly has cactus plants and camels run wild in the desert," he says.

CANADA

ALGERIA

MONGOLIA

BRAZIL

Draw your country's flag.

Explore... explore... explore... (continued)

Look at the list of fictional friends and circle the name of the one who would write the sentences below. [Alasie, Zidan, Rafael, or Batbayar] There is one for each of them. (See the first example.)

Today was colder than usual. The bears are terribly hungry.
[Alasie, Zidan, Rafael, or Batbayar]

Today there was a sandstorm. After that the cactus bloomed.
[Alasie, Zidan, Rafael, or Batbayar]

Today at dawn we saw a family of deer run into a grassy thicket.
[Alasie, Zidan, Rafael, or Batbayar]

Today I taught my parrot to say "Fantastic!" when it sees a wild monkey. [Alasie, Zidan, Rafael, or Batbayar]

A Sets Game: Ecosystems

On the next page there are photos of sets of three things that include a feature of the landscape, a native plant and a native animal for each of 5 ecosystems - ocean, desert, river, mountain, and tropical forest.

1. Cut out the cards and shuffle them.

2. Place them face down in three neat rows of five cards.

3. Pick up two cards. If they belong to the same set, keep them face up in the same place, until you find the third one.

4. If they don't belong in the same set, turn them over again and try to remember where they are.

5. Once you find the three cards for a complete set, pick them up.

6. Continue until you have picked up all five sets.

Here is an example of a **set** = ocean, seaweed, seahorse

ocean

seaweed

seahorse

desert

cactus

spiny lizard

river

berries

river trout

mountain

pine tree

mountain goat

tropical forest

banana

monkey

Environments and Animals

Draw a line to match the biome and the description of a kind of animal that would be a native there. (One example is drawn in red.)

Aquatic	skin that can tolerate hot sunlight
Tundra	limbs to climb or wings to fly
Rain Forest	teeth to chew grass
Grassland	thick fur to keep warm
Desert	smooth waterproof skin

What biome you think this belongs in?

Imagine you could raise a new breed of dog.

Think about the habitat this animal can be comfortable in. Would it need to keep warm with a thick coat of fur or have webbed feet for swimming? Draw this dog in his habitat.

Statement Chart: Biomes

Complete the chart below to sort information about deserts, forests, and tundra. The circles overlap to show what these biomes have in common.

After each statement put an X in the correct column marked D for Desert, F for Forest and T for Tundra.

Check your answers on the next page.

#	Statement	D	F	T
1	This area stays dry with little or no rain.			
2	This kind of biome can have mountains and canyons			
3	This place has a layer of ice that never melts.			
4	Many varieties of plants and animals live here. The ground is fertile and there is enough water.		X	
5	The ground here isn't fertile and it stays dry. The temperature drops at night.			
6	The sun doesn't rise above the horizon for some months of darkness.			
7	All these places have some native animals and plants in an ecosystem.			
8	The moderate climate makes this the easiest biome to make a home.			
9	Harsh climates limit what can survive.	X		X
10	These biomes can exist side by side.			
11	Families don't worry about too much exposure to the sun in this area.			

Answers - Statement Chart: Biomes

Write the number of the statement on the Venn Diagram in the correct place. Does it go in one circle or where the circles overlap? Use each number once to fill all the blank circles.

#	Statement	D	F	T
1	This area stays dry with little or no rain.	x		
2	This kind of biome can have mountains and canyons	x	x	
3	This place has a layer of ice that never melts.			x
4	Many varieties of plants and animals live here. The ground is fertile and there is enough water.		x	
5	The ground here isn't fertile and it stays dry. The temperature drops at night.	x		
6	The sun doesn't rise above the horizon for some months of darkness.			x
7	All these places have some native animals and plants in an ecosystem.	x	x	x
8	The moderate climate makes this the easiest biome to make a home.		x	
9	Harsh climates limit what can survive in this area.	x		x
10	These two biomes can exist very close to each other.	x	x	
11	Families are not very worried about too much exposure to the sun in these places.		x	x

Venn Diagram: Biomes

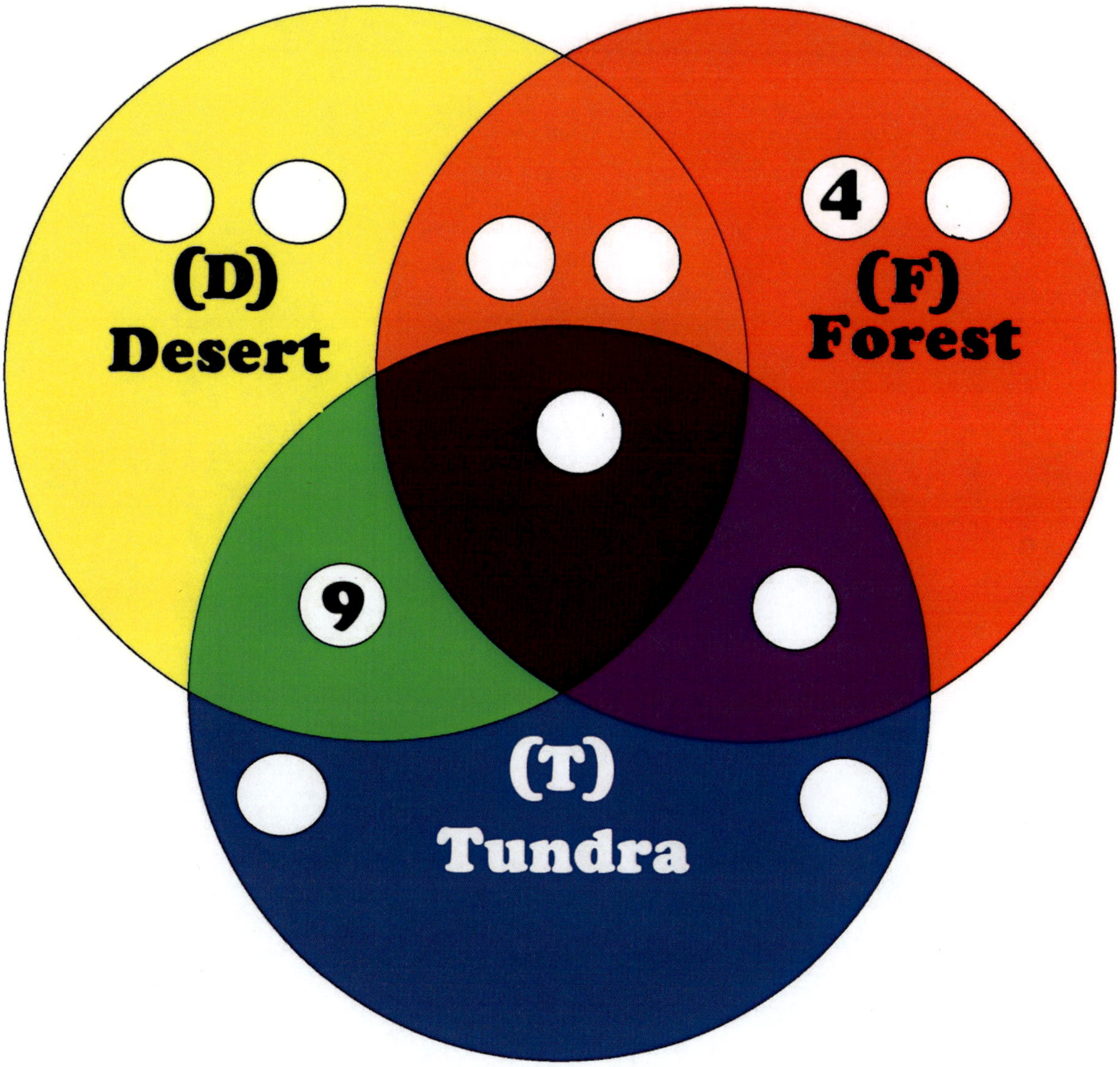

The answers are on the next page. First try to do it yourself.

Answers to Venn Diagram: Biomes

Thinking is more fun if you have someone to listen to your questions and share your thoughts!

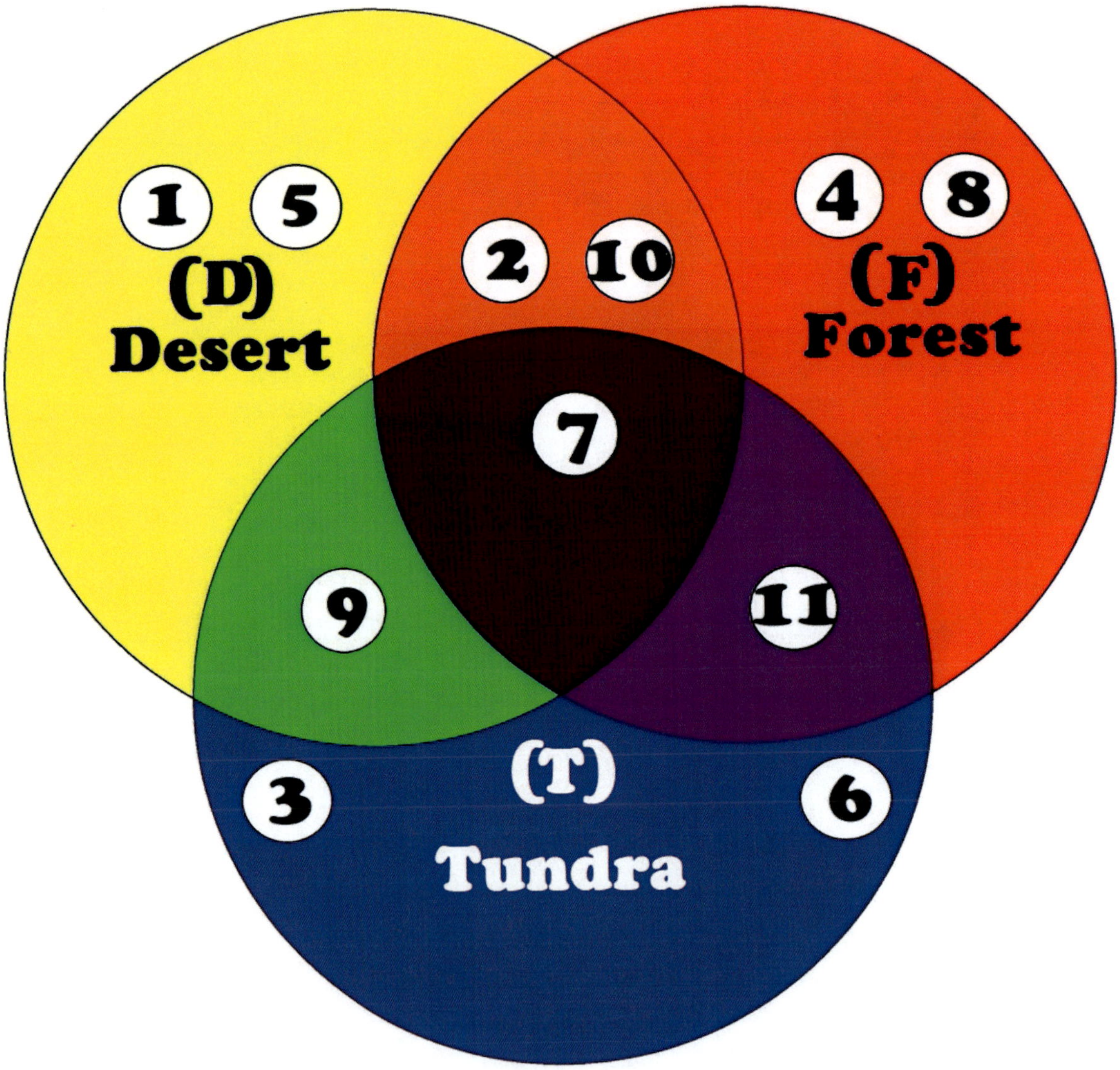

Remember... remember... remember...

The recipes your family cooks depend on your culture and what tastes good to you. Only humans make so many choices about the kinds of food they prefer!

Fill in your preferences below for each main meal.

My favorite breakfast is ______________________.

My favorite lunch is ______________________.

My favorite dinner is ______________________.

What is a food chain?

Where does the story of food begin? The diagram below shows the links from when the sun gives energy to a plant up to the time that a human family has dinner.

All living things are connected by links in a food chain. A vegan diet only includes plants, but most people eat eggs, milk or other foods linked to animals.

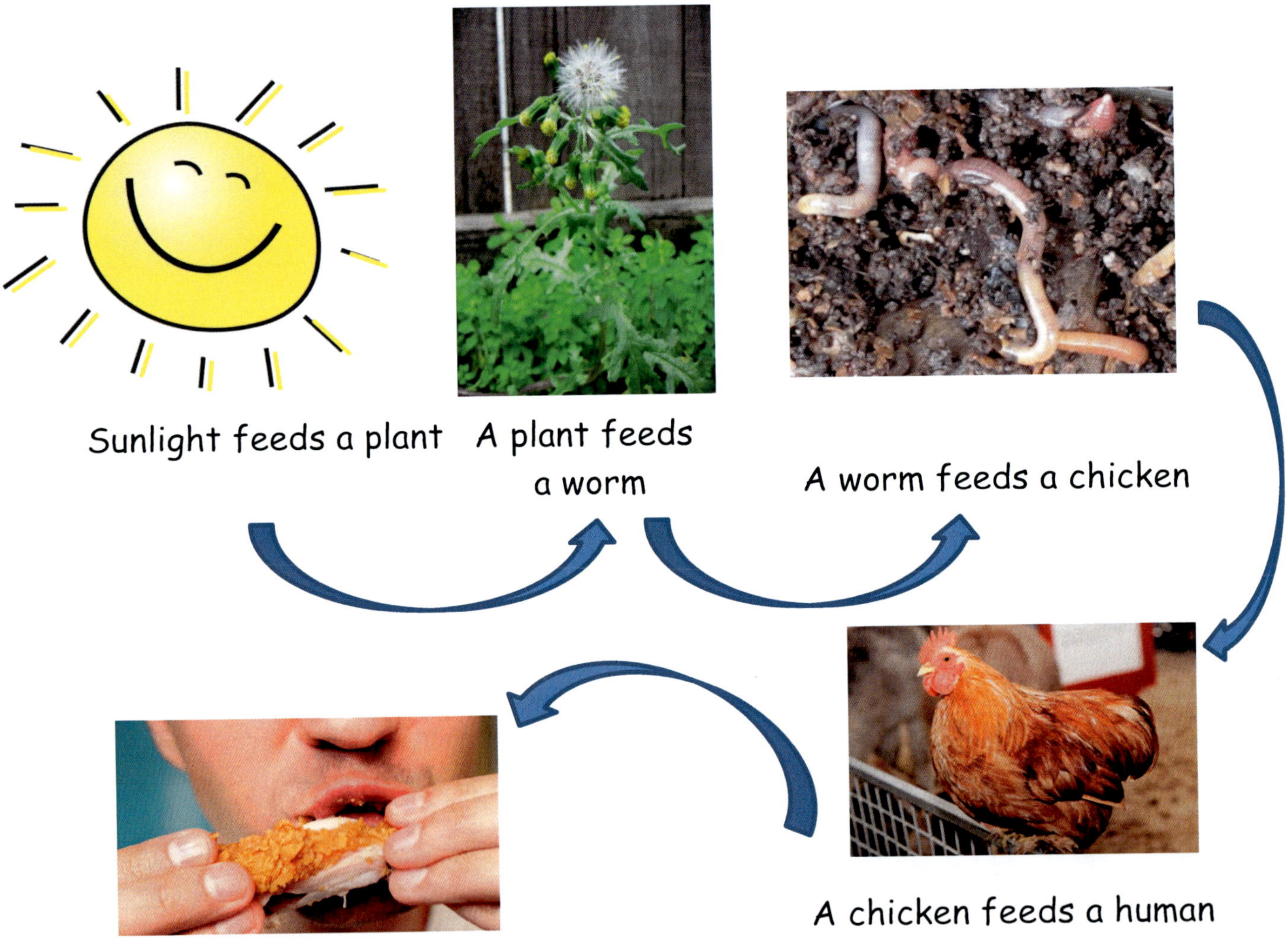

Special Diets

The monarch caterpillar must have one kind of food in its ecosystem. Look at pages 6 and 13 of "The Milkweed, the Monarch and the Moon" (MMM) book to read about this plant.... There are varieties of the milkweed all around the world.

Another special diet is required by a koala bear. The main food of a koala is eucalyptus leaves. Just like the monarch, the koala eats a plant that is toxic to most other animals! Do you know where koala bears are found?

Here are some clues - The koala's native home starts with an A and ends with an A. It is not Asia, which is a different continent, or Austria, which is a country in Europe. The location is down under the equator and it is surrounded by the Pacific Ocean.

Find the answer on a map and write it below

__ __ __ __ __ __ __ __ __

Remember that food is a basic need for all living things, but even people with special diets choose what tastes good to them!

Consider... consider... consider...

Read the ingredients for the recipes below and consider which one has more links in the food chain.

Write down your answer on the line at the end of this page.

1) Fried Rice
<u>Ingredients</u> - 1 cup of brown rice, 1/2 cup cut cabbage, 1/2 cup green peas, 1/2 cup of cut onions, salt, and seasonings to your taste

2) Chicken Soup
<u>Ingredients</u> - 1 chicken leg, fresh parsley, and dill, 1 large cut-up onion, 1 large carrot, salt, and seasonings to your taste

Clue - one of these recipes is vegan.

The ______________________________ has more links in the food chain.

A whole ecosystem is damaged, if any link is broken!

Answer: The **soup** has more links, because the chicken ate plants.

Survival and Endangered Species

If something goes wrong in an ecosystem, the lives of the natives become endangered.

A word for a specific plant or animal is "species."
Each of the pictures below shows a species that was endangered.
Fortunately, corrective actions were taken to rescue them!

Guess which picture matches each named species.

(Write the number under the picture)

(1) Tiny Corroboree Frog
(2) Bongo
(3) Golden Lion Tamarin
(4) Amur Leopard

A species must reproduce during its lifecycle or it will no longer exist.

Do you know the word for a species that no longer exists?

Here it is in scrambled letters: X T I N E C I

E _ _ _ _ _ _

If you said **extinct**, you're right!

Humans have never been an endangered species. After all the threats to survival, more people are alive today than ever!

In the 2nd section we studied:

- different habitats
- ways to sustain a healthy environment
- environments, ecosystems, and biomes
- the food chain

On the next few pages you will find puzzles that use information you learned. Have fun!

Hidden Words

Find words used in this section that are on the list. Their letters are in a line either horizontally, diagonally, vertically, right-to-left or left-to-right.

biomes

R	E	C	T	P	C	H	A	B	I	T	A	T	N
T	A	E	A	L	A	N	D	S	C	A	P	E	A
U	E	S	D	E	R	E	G	N	A	D	N	E	T
N	T	E	T	R	E	S	F	O	B	I	A	I	U
D	A	I	S	N	P	A	S	O	I	C	R	A	R
R	M	C	U	O	E	L	H	Z	L	H	A	H	E
A	I	E	R	C	I	T	A	U	Q	A	M	E	I
E	L	P	V	D	T	C	T	N	A	H	N	A	N
X	C	S	I	V	A	A	O	E	T	E	T	L	H
T	A	T	V	S	H	E	L	T	E	R	E	T	B
I	N	I	A	H	C	D	O	O	F	A	N	H	E
N	O	B	L	B	E	V	I	T	A	N	A	I	A
C	R	A	I	N	F	O	R	E	S	T	L	Q	C
T	T	D	E	S	E	R	T	R	T	T	P	A	H

SHELTER
ENDANGERED
LANDSCAPE
SPECIES
PLANT
NATURE
CLIMATE
PLANET
ZOO
RAINFOREST
EXTINCT
SURVIVAL
FOOD CHAIN
DESERT
HEALTH
TUNDRA
BEACH
HABITAT
AQUATIC
NATIVE

There are 20... How many did you find?

Crossword Puzzle

Across

2. after the last use of a thing, to use its material for a new purpose
4. a plant or animal that was born and lives in its natural habitat
7. the natural habit of doing routine things at night, instead of during the day
10. a kind of forest biome that has a warm and wet climate
13. the physical features, like mountains or lakes, in the environment along with the ecosystem of a natural habitat
14. the way that plants and animals work together in their natural habitat
15. rich in nutrition (food) for life to begin and grow
17. a link where each organism depends on the next as a source of food
19. dependent on being in water to live, like ocean plants and that animals in lakes
20. features of an environment that exist without any effort to make changes

Down

1. a sandy area alongside of an aquatic biome
3. conditions that are in the atmosphere (air) such as temperature, moisture, clouds, and wind speed
5. the view of nature that surrounds and area, like sand dunes or mountains
6. all the natural conditions of an area, including temperate, moisture, and other features, like rivers
8. a natural habitat with the feature of woody plants, like pine or oak trees
9. a man-made shelter for animals to live outside of their native habitat
11. facing threats against well-being with success to stay healthy
12. a biome with a dry climate
16. a species that did live in times past, but no longer does, because it didn't survive to reproduce
18. a biome which is so cold and lacks fertile soil, that hardly anything grows on it

Answers to the Crossword Puzzle

Across

2. after the last use of a thing, to use its material for a new purpose (**recycle**)
4. a plant or animal that was born and lives in its natural habitat (**native**)
7. the natural habit of doing routine things at night, instead of during the day (**nocturnal**)
10. a kind of forest biome that has a warm and wet climate (**rainforest**)
13. the physical features, like mountains or lakes, in the environment along with the ecosystem of a natural habitat (**biome**)
14. the way that plants and animals work together in their natural habitat (**ecosystem**)
15. rich in nutrition (food) for life to begin and grow (**fertile**)
17. a link where each organism depends on the next as a source of food (**foodchain**)
19. dependent on being in water to live, like ocean plants and that animals in lakes (**aquatic**)
20. features of an environment that exist without any effort to make changes (**nature**)

Down

1. a sandy area alongside of an aquatic biome (**beach**)
3. conditions that are in the atmosphere (air) such as temperature, moisture, clouds, and wind speed (**climate**)
5. the view of nature that surrounds and area, like sand dunes or mountains (**landscape**)
6. all the natural conditions of an area, including temperate, moisture, and other features, like rivers (**environment**)
8. a natural habitat with the feature of woody plants, like pine or oak trees (**forest**)
9. a man-made shelter for animals to live outside of their native habitat (**zoo**)
11. facing threats against well-being with success to stay healthy (**survival**)
12. a biome with a dry climate (**desert**)
16. a species that did live in times past, but no longer does, because it didn't survive to reproduce (**extinct**)
18. a biome which is so cold and lacks fertile soil, that hardly anything grows on it (**tundra**)

Here's a true story that's related to a topic we studied:

A team of explorers from America decided to visit the natives of a rain forest in Ecuador. Those natives were a tribe of aborigines and had no contact with other communities in the world.

There was danger because if the tribe was afraid of strangers, they could hurt them. The team believed that life is short anyway, so the risk of danger was worth taking to hope that the future would be brighter if they became friends.

To find out more about what happened, you can read the biographies of Jim Eliot or Nate Saint.

You have reached the end of Section 2!

The Beginning of Section 3

Section 1 was about doing things. Section 2 was about where on earth you may do things, and now Section 3 will focus on thinking about how things happen. We live by patterns!

Think... think... think...

Your MMM book has a hidden message about patterns. Whether you observe the lifecycle of a monarch or the phases of the moon, you can learn about things that are repeated in the same order, time after time.

The value of this book depends on how much you gain awareness of how you use time. Another word for that is being conscientious.

When you pay conscientious attention to something, you can be sure to do your best. In any case, time marches on!

Observe... observe... observe...

The MMM book was inspired by the observance of the moon by a three-year-old boy named Sam. One night, Sam observed that the moon was a complete full circle. He joyfully exclaimed, "Look! The moon is fixed!"

The moon never was alive, yet since the beginning of time, it has worked to measure the passage of time. There is a "new" moon every twenty-nine and a half (29 $\frac{1}{2}$) days. Further, the moon has a pattern of shapes (phases) that marks the passage of the weeks.

These phases are shown below. Every one of our imaginary friends could make a plan to meet on the 15th of the month by watching for the full moon (phase 5). When the first light appears, it will become the first quarter in one week. The week after, less light will appear until the last quarter can be seen (phase 7), and the lunar month ends when there is no light at all shown on the face of the moon (back at phase 1).

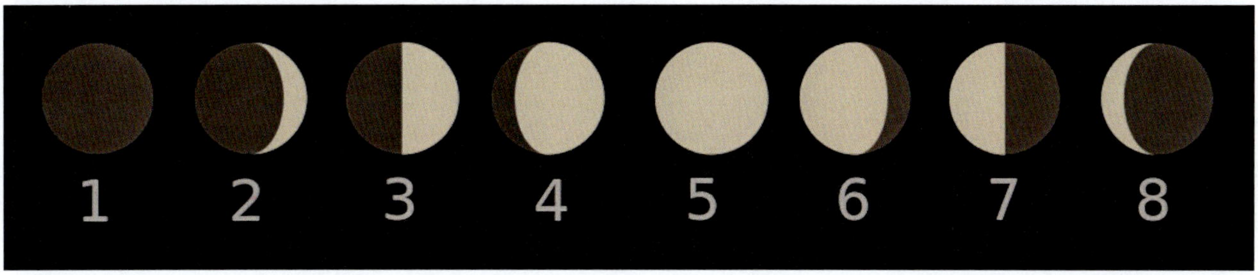

Is the moon ever *actually* completely dark? No... The earth and the moon turn together so that the same side of the moon always faces us. When we see a full moon, the side we don't see is dark, and vice-versa. In other words, when the side we see is dark, the side we don't see has light on it. Vice-versa means reverse, like opposite.

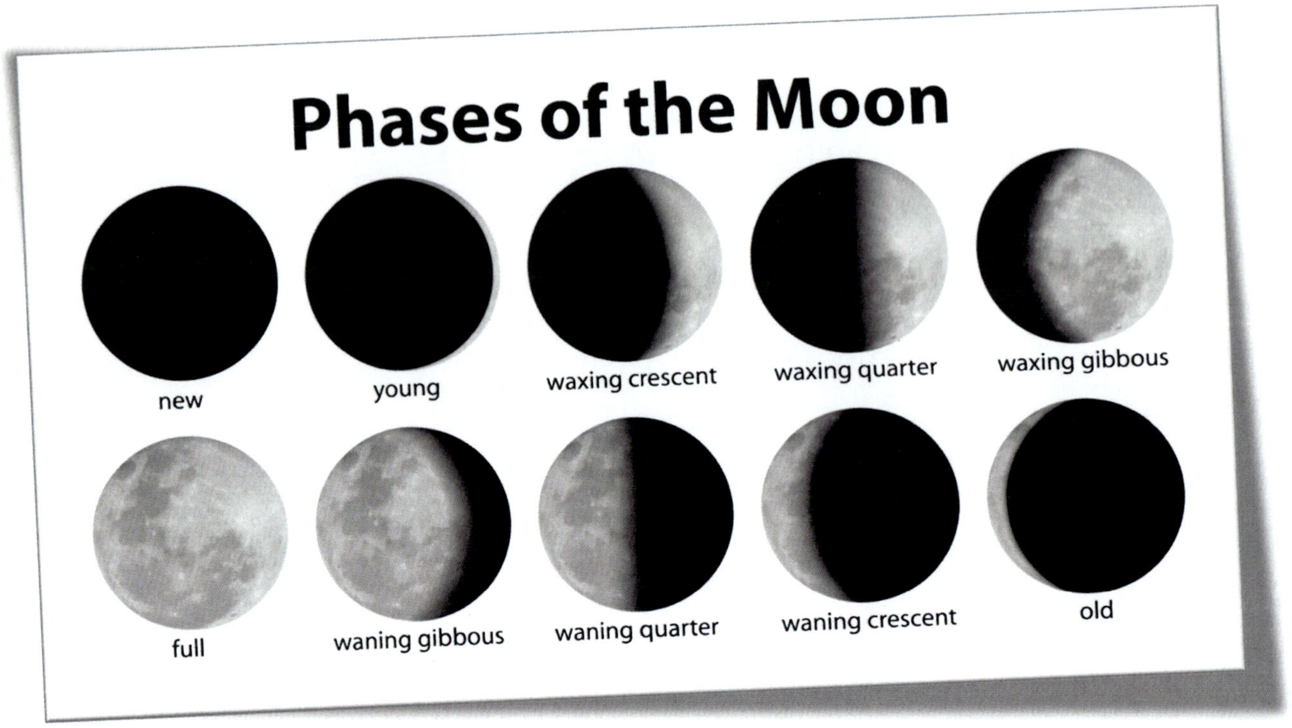

The chart below shows that a new lunar month began on July 1. A week later, on July 8, the moon phase is the first quarter.

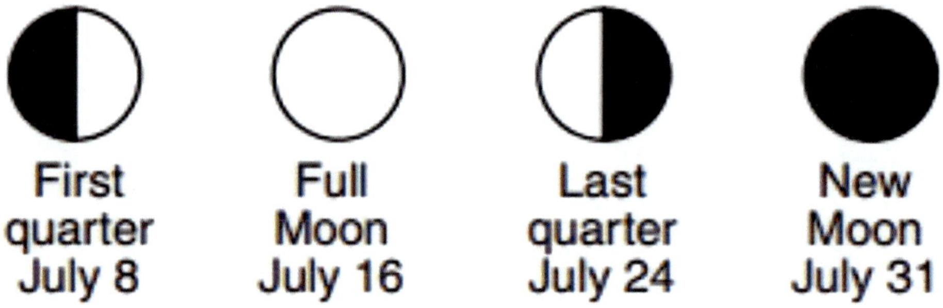

Use the chart above to write the answer in the blank after each question.

If it is July 8, what will the date be when the moon is full in 1 week? _____
...at the last quarter in 2 weeks? _____
...at the end of the lunar month in three weeks? _____

Think... think... think...

What you do in a regular day?

You get up, have meals, and do other things in the same usual order through the week. This is called a following a routine. Like patterns in nature, these things are predictable.

To be healthy you should have time both for routine things and time that you are free to do whatever you like. In this section you can think about how to use all your time in the best possible way.

Here are some terms for three basic ways you spend time -

Free Time is when you do what you like. You may choose to achieve something, but that is up to you. It is your free time. Maybe you can volunteer to help your community... or to try a new a skill.

Self-Care is when you show how grown up you are. For example, you can manage on your own to be washed and dressed, get enough sleep, exercise, and eat the right food.

Tasks get done in routine times when you know what people expect you to do. For example, you are responsible for doing your chores and homework. These things can require skill, teamwork, and perseverance to do a job well.

If you do a thing for 2 weeks, that forms a habit. Complete the sentence below:

One good idea I have to form a habit is that every day I will...

____.

Put a check mark on the picture of a routine activity you do and an x on a free time activity that you would like to do.

Remember... remember... remember...

Mark the sentences below with a T for **true** or an F for **false**.

[ ] Milkweed leaves are healthy for monarch caterpillars, but poison for most other insects.

[ ] When a monarch egg hatches-- a winged insect flies out.

[ ] In the four seasons, after the winter comes summer.

[ ] There is a full moon on the 15th day of the lunar month.

[T] Milkweed plants are poisonous to many animals and humans, but not to monarch caterpillars

[F] A caterpillar crawls out of the egg when it hatches

[F] After the winter comes spring

[T] A lunar month is 29 days with a full moon on Day 15

A Sets Game: Projects

This memory matching game has three cards for a set about 5 kinds of projects -

- bake cookies
- practice music
- plant a garden
- conduct an experiment
- take a trip

1. Cut out the cards and shuffle them.

2. Place them face down in three neat rows of five cards.

3. Pick up two cards. If they belong to the same set, keep them face up in the same place, until you find the third one.

4. If they don't belong in the same set, turn them over again and try to remember where they are.

5. Once you find the three cards for a complete set, pick them up.

6. Continue until you have picked up all five sets.

Here is an example of a set:

Beginning

Middle

End

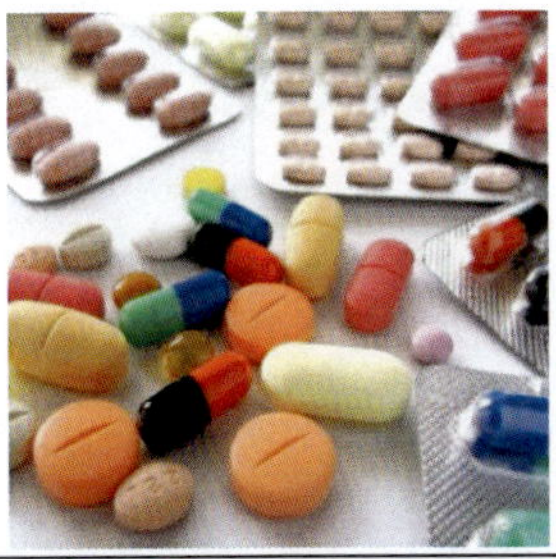		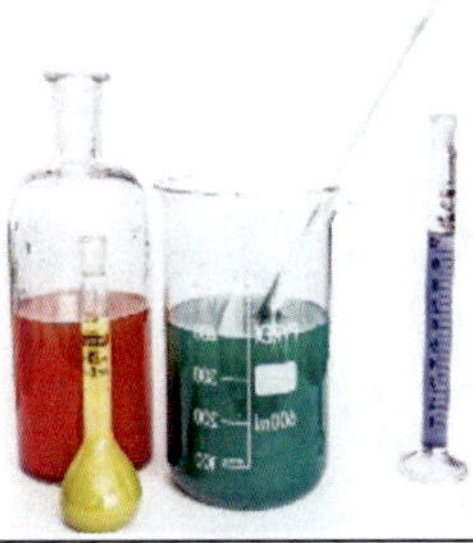

Here is a chart of words about time and clues to their meanings:

Words	Clues to meanings
alarm clock	a clock that can wake you up
balance	arrangement so things move easily
calendar	a chart with months, weeks and dates
challenge	a goal with obstacles to overcome
checklist	things to be marked off when done
clock	a tool to measure time
discipline	strength to focus on doing something
focus	concentrate against distractions
goal	something to achieve or arrive at
hour	a unit of time equal to sixty minutes
hourglass	a tool to observe the time by how much sand falls in a container
minute	a unit of time measured by counting sixty seconds
practice	to do something over and over again so that you can do it better
sacrifice	to give up one thing to get another thing
schedule - noun	a time set for something to happen... for example when a bus will come
schedule - verb	to set a time for something to happen, like going to the dentist
second (not 2nd, the sequence after 1st)	a unit of time that is just long enough to say "one thousand"
sundial	a tool to observe the time of day by a shadow
twenty-four seven	all hours of all days in the week

Think... think... think...

Observing time...

Mark each way that you can observe time passing with an [X].

[ ] a green sprout shoots up from where you planted a seed
[ ] a baby learns to walk
[ ] the stars shine all night
[ ] another candle is added to your birthday cake
[ ] the phases of the moon repeat their pattern
[ ] the calendar is turned to a new month
[ ] snow stays on the peak of a mountain

Answer: Stars and snow that do not show any change are <u>not</u> ways to see time was spent. The other events show time has passed.

Spending Time

Write an answer on the line after each question below. Pictures are given to suggest an idea. You could also add a drawing, if you like.

What can you do in a minute? ______________________________

What can you do in an hour? ______________________________

What can you do in a day? ______________________________

Spending Free Time

Put an X beside ways listed below that you spend your free time

[ ] I play games on my phone
[ ] I have a hobby, which is ____________________
[ ] I play with my friends
[ ] I like to be alone
[ ] Reading or writing
[ ] Watching entertainment
[ ] Volunteering to help on a community project

Tell about another activity that you would like to do on the lines below.

____

____

____

____

Time flies!

Usually, there doesn't seem to be enough time to do everything we wish. With skills to plan to do what is most important, and some sacrifice of less important things, you can reach a goal!

Everyone feels more comfortable around people who have good manners. So... when would it be proper behavior to put your hands on your face and to point? Only if you were a clock! Let's spend some time getting clear about clocks!

Clocks that use hands that move round on a dial or clockface are called 'analogue' clocks (pronounced - 'Anna-log'). They usually have two moving pointers called 'hands'.

- a shorter hour hand that takes **1 hour** to move from one number to the next. It goes all the way round the clockface in 12 hours (half a day).

- a longer minute hand that takes **5 minutes** to move from one number to the next. It goes all the way round the clock face in exactly one hour.

Sometimes these clocks have three hands - an even longer hand, the second hand, that takes only **5 seconds** to move from one number to the next, and sweeps around the whole clockface in just one minute.

Think... think... think...

When the minute hand points straight up to the **12** at the top, this is the start of an hour.

Look at the hour hand to see which hour is starting.

The picture above shows it is **4 o'clock**.

The minute hand is straight up to the **12** at the top, so it's the start of an hour.

Which hour? Look at the hour hand - it is pointing to the **4**, so it is exactly **4 o'clock** and no extra minutes.

Draw... draw... draw...

Here are some clocks without hands. Draw in the hands to show the correct times.

Remember the minute hand is **long**, and the hour hand is **short**.

Very useful tip: *To be exact and accurate, use a ruler to draw your straight lines.*

2:30

3:15

5:20

6:05

12:00

10:45

Ask an adult to help you check your answers!

These analogue clocks are very clever, because they use the same circle of numbers to show seconds, minutes and hours passing accurately, so you can tell the time at a glance.

In between the numbers, there are normally 5 small marks at the rim of the clockface, so you can see that as the minute hand reaches each of these marks, a full minute of time has passed.

When the minute hand moves from **12** to **1**, it has passed 5 small minute marks – 5 minutes' time has passed.

When it reaches **2**, another 5 minutes has passed. Another five minutes takes it to **3**, and so on.

The minute hand moves right round the clockface in **1 hour.**

If you were to look carefully and patiently at a working analogue clock, you would notice that when the minute hand has moved all the way round the clockface once, the hour hand has moved to the next number, and the fastest moving second hand will have travelled round the clockface sixty times.

You might be dizzy, having watched the speedy second hand rotate sixty times, but you will have observed that....

60 seconds = 1 minute
60 minutes = 1 hour

Well, usually you don't have to think about the seconds, when you're telling the time. So, let the clock do the counting...... you have more important things to do!

Let's look at the math!

There are **12** numbers on a regular analogue clockface.

Each number is **5 minute-marks** away from the number before.

That's **12** multiplied by **5**. The first step in learning how to multiply is understanding repeated addition. That means...

12 + 12 + 12 + 12 + 12 = **60**

Is the same way of saying...

12 X 5 = **60**

There are **60** **minutes** in each **hour**.

If we know there are **60** minutes in one hour, we can work out that there are **30** minutes in half an hour, because...

half of 60 is 30

and when the minute hand is pointing straight down to the bottom of the clockface, it is halfway around, which means **6** times (X) **5** minutes have passed...

6 X 5 = 30

Draw... draw... draw...

Here are some more blank clockfaces for you to complete, but this time you will be showing times at half-past the hour.

In each case, the minute hand will be pointing straight down to the **6** at the bottom of the clockface...

But the hour hand will have moved halfway to the next number...

It's even more important to use a ruler to be really accurate!

The first one has been done for you.

4:30

7:30

10:30

1:30

12:30

6:30

So far, we've looked at *o'clock* and *half-past*. The minute hand was at **12** for *o'clock*, and at **6** for *half-past*.

Now we'll add quarter after, and quarter to.

At...

- o'clock, the minute hand points up to 12
- half past, the minute hand points down to 6
- quarter after, the minute hand points middle right to 3
- quarter to, the minute hand points middle left to 9

Don't forget the hour hand also moves.

At quarter after, it has moved a quarter of the way to the next number.

At quarter to, it has moved till it has only a quarter the distance still to go to the next number.

You will get used to judging these distances.

Smaller Divisions for the Minute Hand

Can you work out the time on these clocks ?

Remember...

- When the minute hand has travelled more than halfway round the clockface, it is correct to tell how many minutes are left before it reaches the top again.
- Count the small marks between the numbers to know exactly how many minutes past the hour, or minutes till the next hour.

The first two have been done to show you how this works.

10 minutes past 5

5:10

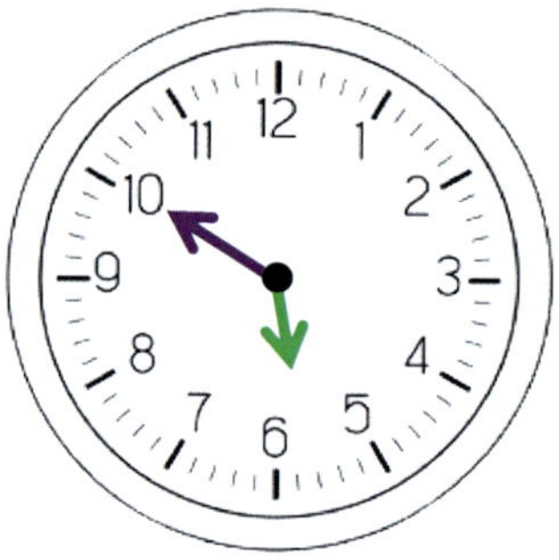

10 minutes to 6

5:50

20 minutes past 2

______

25 minutes past 9

______

5 minutes to 4

______

20 minutes to 12

______

Digital Clocks

All of this time-telling has been about analogue clocks, but there are many clocks that don't have any hands.

These are called digital clocks, because they just show the time in numbers that change as the minutes tick by.

17-minutes past 2

2-minutes past 6

1-minute past 4

52-minutes past 8, which is the same as 8-minutes to 9

Did you notice that in digital time, you don't say how many minutes 'to' the next hour, even if there are only a few left?

Digital clocks only ever tell you how many minutes past the hour it is.

The interesting thing about digital clocks is that they are twenty-four-hour tools, not twelve-hour clocks.

So, once you get past **12:59** at lunchtime, the next hour is **13:00**, and so on, until midnight.

Some digital clocks show one-minute past midnight as **24:01**, but most show it as **00:01**.

Busy schedules need a plan!

On this page are two scenarios where time needs to be measured carefully. See if you can advise Abigail and Sam.

Scenario 1

Abigail has **one whole hour** of free time before she must do her chores.
The walk to her friend's house takes **a quarter of an hour** (15 minutes).
The game they want to play takes **a half an hour** (30 minutes).
Her walk home also requires **a quarter of an hour** (15 minutes).
Is **1 hour** enough time to visit her friend and play the game?
Circle the correct answer: YES or NO

Scenario 2

Sam has **one free hour** between 3 o'clock and 4 o'clock.
He wants to meet a friend who lives **a quarter of an hour** (15 minutes) away.
His mom wants to hear him practice piano for **half an hour** (30 minutes).
His mom comes home at **a quarter past 3** (15 minutes after).
Sam wants to go outside at **a quarter to 4** (15 minutes before).
Does he have enough time to fit in all that he wants?
Circle the correct answer: YES or NO

The answer for both scenarios is YES! If Abigail and Sam keep track of time, it is possible to follow their plans.

Think about... Think about... think about...

Everything you do takes time...

Use the word bank below to help fill in the empty lines and write about something you would like to do.

Word Bank:

drawing, singing, playing a musical instrument, dancing, swimming, sports, cooking, doing puzzles, math problems, reading, writing, exploring my environment, building things, fixing things, taking care of animals, volunteering in my community, gardening

When I have free time it might be nice to try...

____

Here are quotes from successful people:

- "Hard work beats talent, if talent doesn't work hard."
  - Tim Notke

- "Genius is 1% inspiration and 99% perspiration."
  -Thomas Edison

  (perspiration means sweat)

- "Without labor - nothing prospers." – Sophocles

- "I'm a great believer in luck, and I find the harder I work the more I have of it." - Thomas Jefferson

- "The only place where success comes before work is in the dictionary." - Vidal Sassoon (think about alphabetical order)

Write about something you might succeed at if you work harder:

____

____

____

Here are descriptions of ways to spend time. Use the statement chart to sort out what applies to these ways.

Free Time - This time is to do what you please. For example, you might choose to volunteer or practice a skill.

Self-Care - Time to make sure you sleep enough, eat nourishing food, wash, dress, and other things you do to stay well.

Tasks - Time to be responsible for the things that you are expected to do, like take out the trash, feed a pet, and do your homework.

Read each statement and put an X under when to do that. These are Free Time (F), time for Self-Care (S), and time we must do our Tasks (T). For example, Statement 2 has an X in column F and S. This statement only applies to free time and self-care.

Statement Chart: Spending Time

#	Statement	F	S	T
1	You can think about doing anything you'd like			
2	This is when it is okay to ask others to give you complete privacy	X	X	
3	This is when you should be a team player			
4	Time to brush your teeth and other personal care tasks			
5	You might use this time for a hobby			
6	Show you are responsible to cooperate with others			
7	Keep track of the time you are spending			
8	Keeping a journal may become part of your routine			
9	Check at the library for a community service to do			
10	Manage to take care of what's important to you first			
11	Look for ways to get routine things done more easily			

Answers for Statement Chart: Spending Time

#	Statement	F	S	T
1	You can think about doing anything you'd like	X		
2	This is when it is okay to ask others to give you complete privacy	X	X	
3	This is when you should be a team player			X
4	Time to brush your teeth and other personal care tasks		X	
5	You might use this time for a hobby	X		
6	Show you are responsible to cooperate with others			X
7	Keep track of the time you are spending	X	X	X
8	Keeping a journal may become part of your routine	X	X	
9	Check at the library for a community service to do	X		X
10	Manage to take care of what's important to you first	X	X	
11	Look for ways to get routine things done more easily		X	X

Venn Diagram: Spending Time

Fill in the empty white circles with your answers from the chart on the previous page. One circle has been filled in as an example.

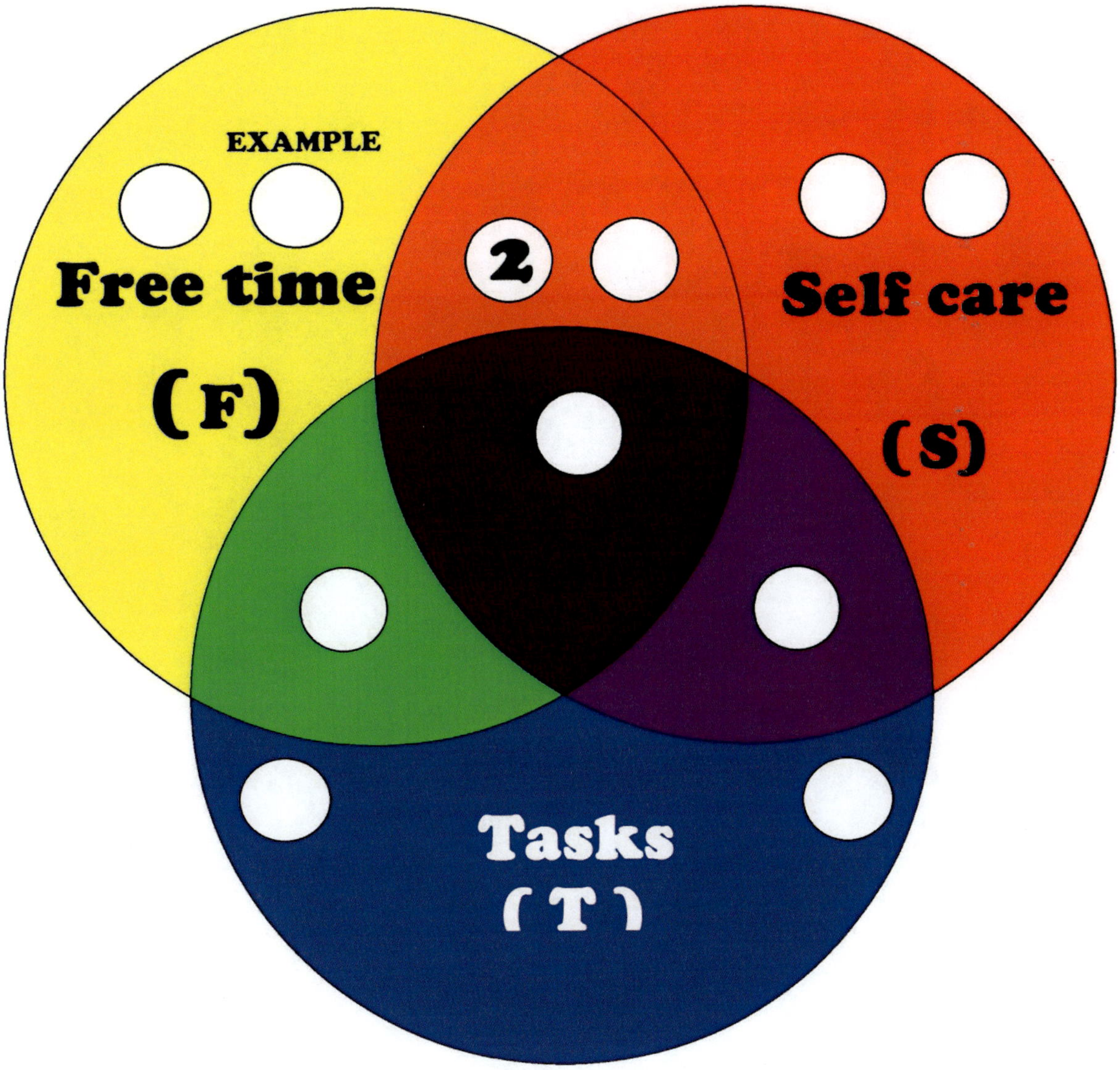

Answers for Venn Diagram: Spending Time

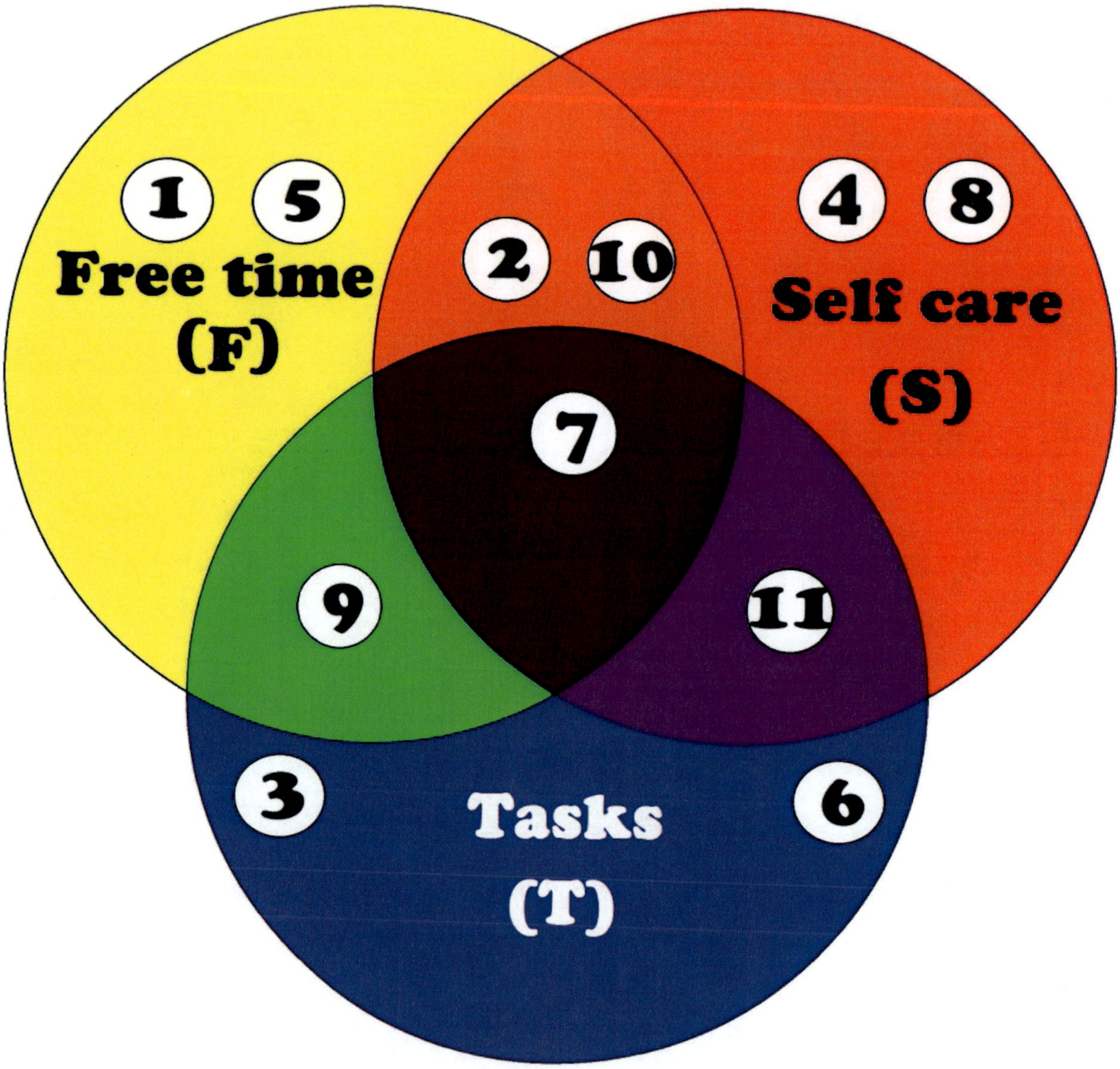

Mystery Word

This code uses symbols for valuable abilities, such as to listen, to focus on a target, to practice for a performance, and to use technology.

 = I

 = K

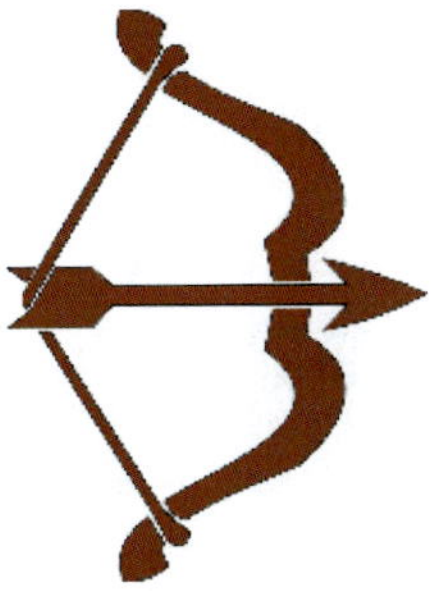 = S

 = L

The picture code spells the mystery word. Use the last letter twice.

_____ _____ _____ _____ _____

The answer is underlined in the sentence below.

You can try to find time to practice and gain a new <u>skill</u>!

Consider... consider... consider...

Fill in the boxes with the number of hours you spend doing each activity on a routine day.

Also fill in the blank spaces to complete the sentences.

[ ] Eating
[ ] Sleeping
[ ] Self-care and exercise, such as ______________________________ .
[ ] Attending school
[ ] Routine chores, such as ______________________________________
[ ] Studying and practicing skills, such as ________________________
[ ] Hobbies, such as ____
[ ] Citizenship activities, such as ______________________________
[ ] Entertainment, such as _______________________________________

Total of 24 Hours = 1 Day

Remember: your total must include at least 7 hours of sleep!

Use a separate paper to work out your total. This exercise can help you to consider how you would like to use your time in the future.

In the 3rd section we studied:

Some things we learned about were:

- ways to tell time
- ways to use your time

Some of the skills you practiced were:

- reading the face of a clock
- scheduling activities

On the next few pages you will find puzzles that use information you learned. Have fun and good luck!

Hidden Words

Try to find the **20 words** on the list that are related to the subject of the lunar month. They are hidden horizontally, diagonally, or vertically. The letters may be backwards.

lunar month

R	M	F	I	R	S	T	Q	U	A	R	T	E	R
N	E	T	I	M	E	M	H	V	N	T	W	E	P
T	T	K	E	E	W	O	T	T	E	H	N	R	H
S	T	I	B	R	O	O	U	H	W	A	R	R	A
L	U	N	A	R	R	N	T	G	T	L	E	T	S
H	T	R	A	E	E	R	H	I	U	F	T	T	E
L	I	L	I	G	H	T	A	E	P	A	T	D	S
T	F	H	I	C	E	C	R	Y	E	H	A	O	A
U	T	R	E	F	L	E	C	T	R	D	P	B	R
S	E	L	M	O	N	T	H	N	A	V	E	S	E
R	E	V	O	L	V	E	E	E	N	U	S	E	I
Q	S	S	I	N	P	E	V	W	S	L	I	R	S
E	D	I	S	K	R	A	D	T	Y	S	O	V	S
S	O	L	A	R	S	Y	S	T	E	M	T	E	R

OBSERVE
EARTH
LIGHT
SOLARSYSTEM
NEW
FIRST QUARTER
HALF
TWENTY EIGHT
TIME
DARKSIDE
PHASES
REVOLVE
MONTH
REFLECT
MOON
WEEK
SUN
LUNAR
PATTERN
ORBIT

Crossword Puzzle

Solve the clues below and fill in the puzzle. Try to figure out the answers yourself before looking at the answers on the next page.

Across

3. a repeating path that one object in space travels around another one

5. the gases surrounding the earth or another planet

7. the sun and all the objects that orbit around it (two words)

10. to watch carefully and see what happens

11. a period of time made up of 7 days

Down

1. the different ways the moon looks from earth (two words)

2. the first lunar phase (two words)

4. the absence of light

6. one of the names for our planet

8. what the moon reflects from the sun that makes it shine

9. shapes or numbers repeated in the same order

Answers to the Crossword Puzzle

Across

3. a repeating path that one object in space travels around another one (**orbit**)

5. the gases surrounding the earth or another planet (**atmosphere**)

7. the sun and all the objects that orbit around it (two words) (**solarsystem**)

10. to watch carefully and see what happens (**observe**)

11. a period of time made up of 7 days (**week**)

Down

1. the different ways the moon looks from earth (two words) (**lunarphases**)

2. the first lunar phase (two words) (**newmoon**)

4. the absence of light (**dark**)

6. one of the names for our planet (**earth**)

8. what the moon reflects from the sun that makes it shine (**light**)

9. shapes or numbers repeated in the same order (**sequence**)

Hidden Words

Try to find the **22 words** on the list that are related to the subject of time. They are hidden horizontally, diagonally, or vertically. The letters may be backwards.

P	H	X	A	T	C	A	E	D	S	R	P	R	D	E
H	P	G	T	L	S	A	G	F	J	C	A	K	I	C
H	O	F	M	R	A	I	L	C	F	A	Z	L	S	I
Z	R	F	J	O	I	R	L	E	O	O	Y	M	C	F
W	A	T	C	H	X	W	M	K	N	A	R	S	I	I
E	C	I	T	C	A	R	P	C	C	D	S	T	P	R
T	O	D	O	L	I	S	T	N	L	E	A	C	L	C
T	W	E	N	T	Y	F	O	U	R	O	H	R	I	A
S	C	H	E	D	U	L	E	G	G	A	C	C	N	S
E	T	U	N	I	M	I	O	O	L	J	H	K	E	T
F	S	M	H	Q	R	R	A	L	E	O	T	S	X	A
O	F	K	N	C	P	L	E	C	N	A	L	A	B	R
C	K	F	I	S	U	N	D	I	A	L	A	L	Y	G
U	S	S	A	L	G	R	U	O	H	H	E	E	O	E
S	O	T	M	E	L	I	T	Y	V	R	H	U	V	T

ALARMCLOCK
BALANCE
CALENDAR
CHALLENGE
CHECKLIST
DISCIPLINE
EFFORT
FOCUS
GOAL
HEALTH
HOURGLASS
MINUTE
PRACTICE
PROGRESS
SACRIFICE
SCHEDULE
SKILL
SUNDIAL
TARGET
TODOLIST
TWENTYFOUR
WATCH

You have reached the end of Section 3!

Take a break to
color this page
before starting the
next section

The Beginning of Section 4

Section 1 sparked your imagination to think about how you can make things better and **Section 2** explored where. **Section 3** looked at when you could find time. Now in **Section 4** let us plan to take action toward a goal... step by step!

This is a picture of Earth from the surface of the Moon.

Think... think... think...

The Milkweed, the Monarch, and the Moon book tells about how the pattern of the moon's phases continues, no matter what you do with that time. There will always be a future!

Quote from a famous baseball player:
"The future ain't what it used to be." -Yogi Berra

History tells stories of extremely difficult times when people fought very hard to survive against disasters. In those times the extra effort to be healthy made communities stronger and smarter.

The future is always bright when people honestly care about each other. As long as you set a good goal and honestly try to reach it, there is hope for improvement.

Put an X after the following reasons you would like to set a goal.

[ ] Setting a goal makes me think about what is possible.
[ ] Working toward a goal makes me grow smarter and stronger.
[ ] Trying to achieve something shows I'm thankful for being alive.
[ ] Working toward a goal makes me a part of a team.
[ ] Success for reaching a goal will bring rewards that I want.

Write about a goal you would like to reach:

____

____

Write a sentence on the lines below about anything that you did that you thought you could never do. For example, did you need patience to learn to ride a bike, to swim, or to make a melody on a piano?

Put something in the frame below to show or tell what you did.

"If you can dream it, you can do it."

-Walt Disney

Use the words below to write about your dream:

set goals

dream

make an action plan

don't give up

Explore... explore... explore...

Plan to bake a batch of oatmeal cookies. If this example is not suitable, use this model to plan another activity. Think of the amount of time, which skills, and what things are needed.

Only operate the oven with adult supervision!

Activity in four steps	How long can it take?	Which skills do I need to use?	What things are needed?
Buy the list of ingredients for the recipe	1 hour	Read the recipe and find items	List the ingredients and find way to carry them
Prepare the kitchen area and tools, preheat the oven	1/2 hour	Only operate the oven with adult supervision	Measuring tools, mixing tools, oven
Follow the directions to put cookies into oven and let them bake	1 hour	Know how to follow steps with care and patience	Baking pan, potholder, cooling rack
Prepare to serve	1/4 hour	Think of how to serve in a neat and healthy manner	Serving dish, knife, plates, forks, napkins, and drinks

Unscramble the letters to make a word and use it to complete the sentence below: R J P O C E T

The answer is in the next paragraph.

When you decide to do something with a plan, you are starting a
P _ _ _ _ _ _

Having a project gives an opportunity to develop skills and character. Choose carefully so that your efforts have satisfying results, but in any case, you will learn from the experience every time you try your best to achieve something!

Quote:

"Failure is simply the opportunity to begin again, this time more intelligently." - Henry Ford

What do you understand from the above quote?

____

____

____

Write... write... write...

Write about something you would like to do and why it is worth doing.

Remember... remember... remember...

Remember something that you failed to complete.

Mark and X by the problem you had. Your experience will help you to succeed next time!

I needed...

[ ] more preparation of skills
[ ] more preparation of materials
[ ] more dedication of time and energy to make an effort
[ ] more teamwork from the right kind of team members

A quote is anonymous when we don't know who said it first...

"If at first you don't succeed, try, try again!" - Anonymous

Pick one of the things you learned about from a failed effort and write about how your next try will be more likely to succeed. Remember not to let blame or any negative thinking make you feel stuck - take courage!

I learned... ____

____

____

____

____

Setting a Goal

Complete the statement chart using the definitions below:

- Skills - specific abilities to do a job
- Good Character Trait - the behavior that is normal for a person
- Material - inanimate object that must be on hand for something to be done

There are three columns to mark if the statement is correct: "S" for Skills, "C" for Character Trait, and "M" for Material Needs.

The statements apply to one, two, or three definitions. Do your best, then have a look at the answer sheet on the next page.

Statement Chart: Setting a Goal

#	Statement	S	C	M
1	This takes professional training and practice			
2	This takes self-discipline to improve			
3	When a goal is almost reached, you need less of this			
4	This matters the most in any teamwork			
5	This determines how well the work will be done	X		
6	You should make sure you have enough of this before you start			
7	Without this, there's not much hope of success			
8	This makes it possible to face any problems with confidence			
9	This can be measured			
10	This should increase even after a goal is reached			
11	This does **not** require professional training to get it			

Answers for Statement Chart: Setting a Goal

#	Statement	S	C	M
1	This takes professional training and practice	X		
2	This takes self-discipline to improve	X	X	
3	When a goal is almost reached, you need less of this			X
4	This matters the most in any teamwork		X	
5	This determines how well the work will be done	X		
6	You should make sure you have enough of this before you start			X
7	Without this, there's not much hope of success	X	X	X
8	This makes it possible to face any problems with confidence		X	
9	This can be measured	X		X
10	This should increase even after a goal is reached	X	X	
11	This does **not** require professional training to get it		X	X

Venn Diagram: Setting a Goal

Fill in the empty white circles with your answers from the chart on the previous page. One circle has been filled in as an example.

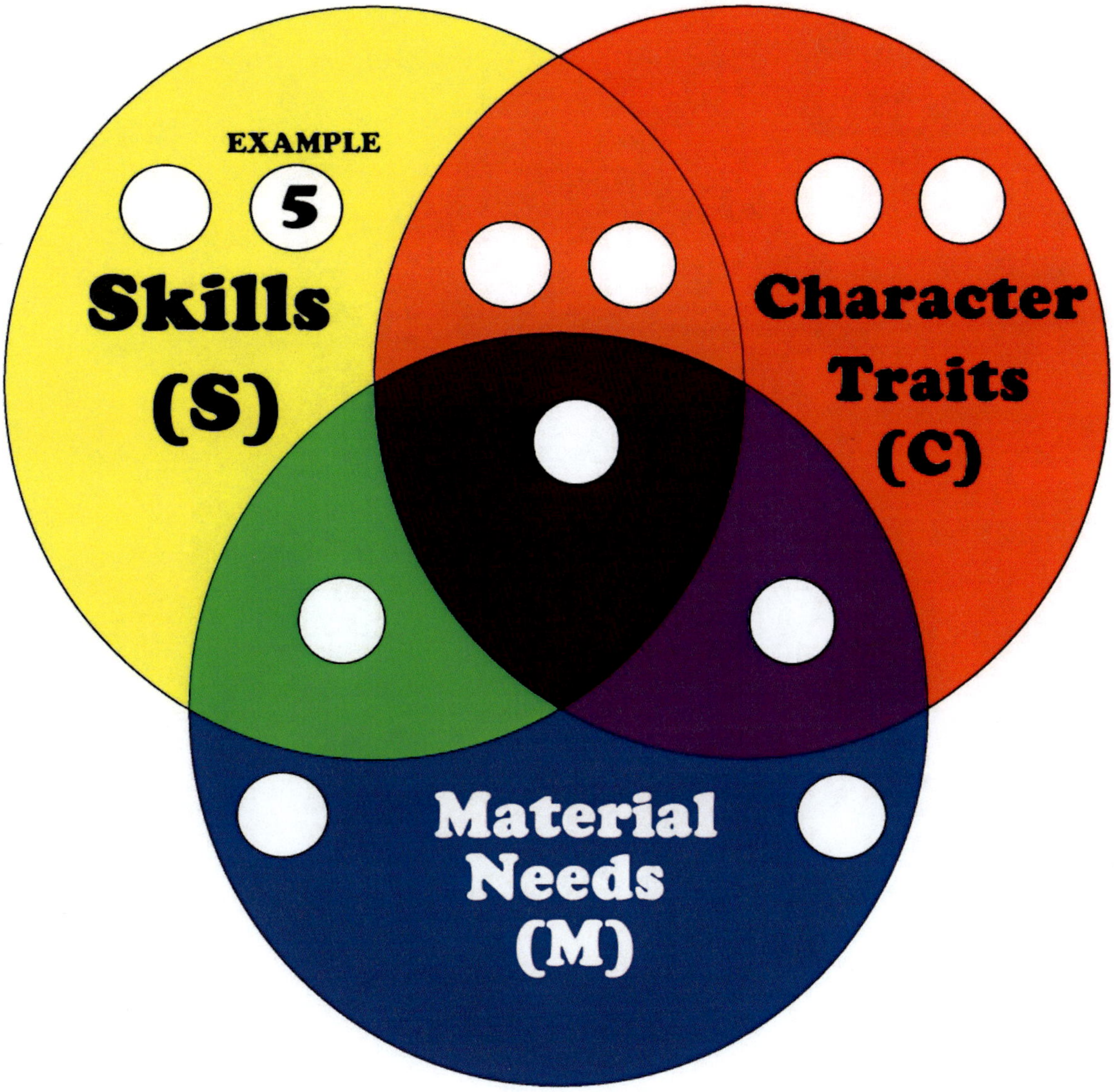

Answers for Venn Diagram: Setting a Goal

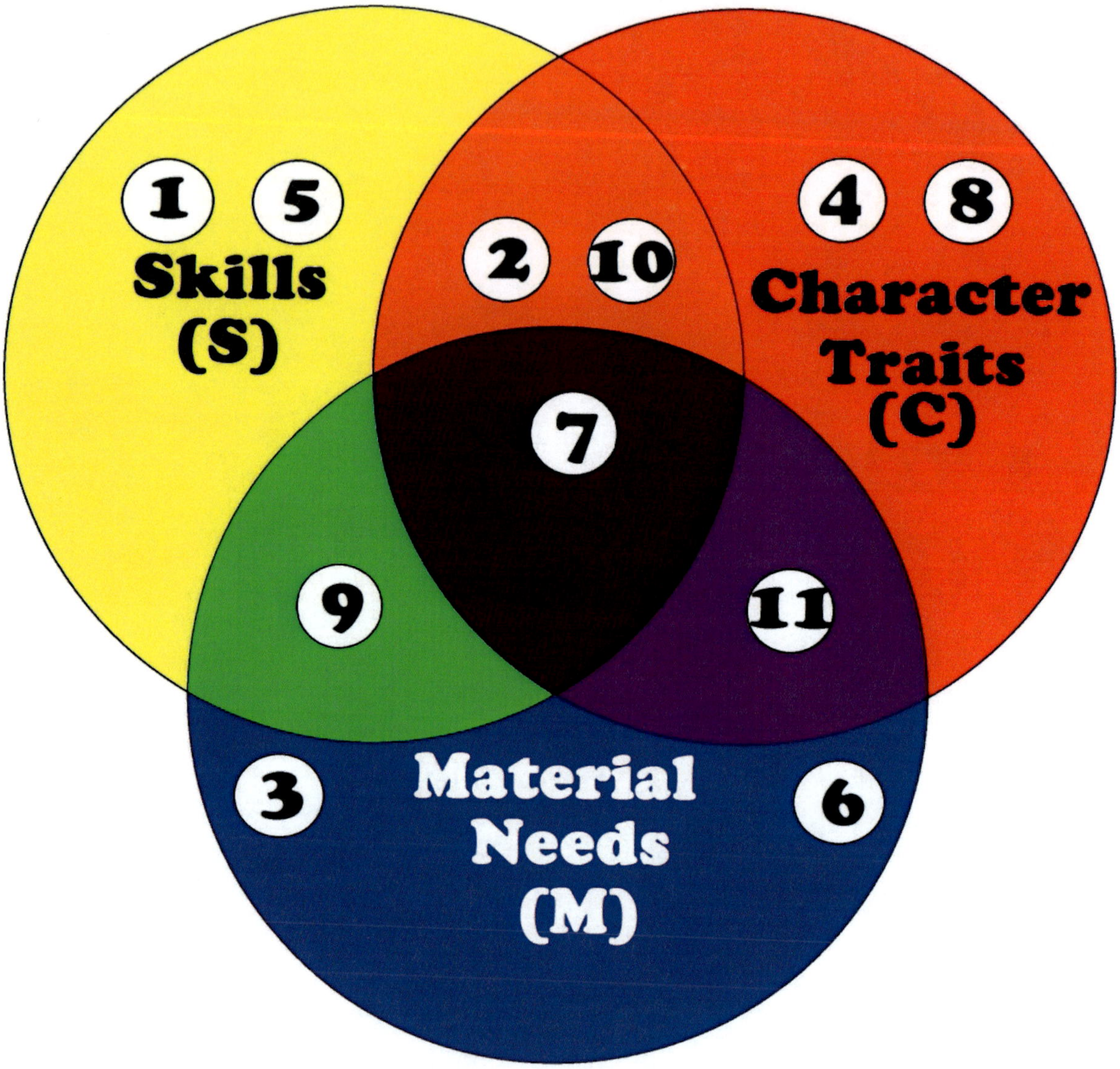

Defining Character

Draw a line to connect the words below with their definition:

character	the ability to keep going at something
temptation	a desire for something at the wrong time
tenacity	a person's regular behavior

Would you pick a team member who has obviously good skills but unclear character or someone with obviously good character with unclear skills? Explain your choice on the lines below.

____

____

____

True or False: Defining Character

Fill in the box with T for True or F for False after you think about each statement.

[ ] The most important factor for success is skill.
[ ] Temptations and distractions are good reasons to give up.
[ ] A good character trait that successful people have is tenacity.
[ ] There is no way to predict who is likely to succeed.

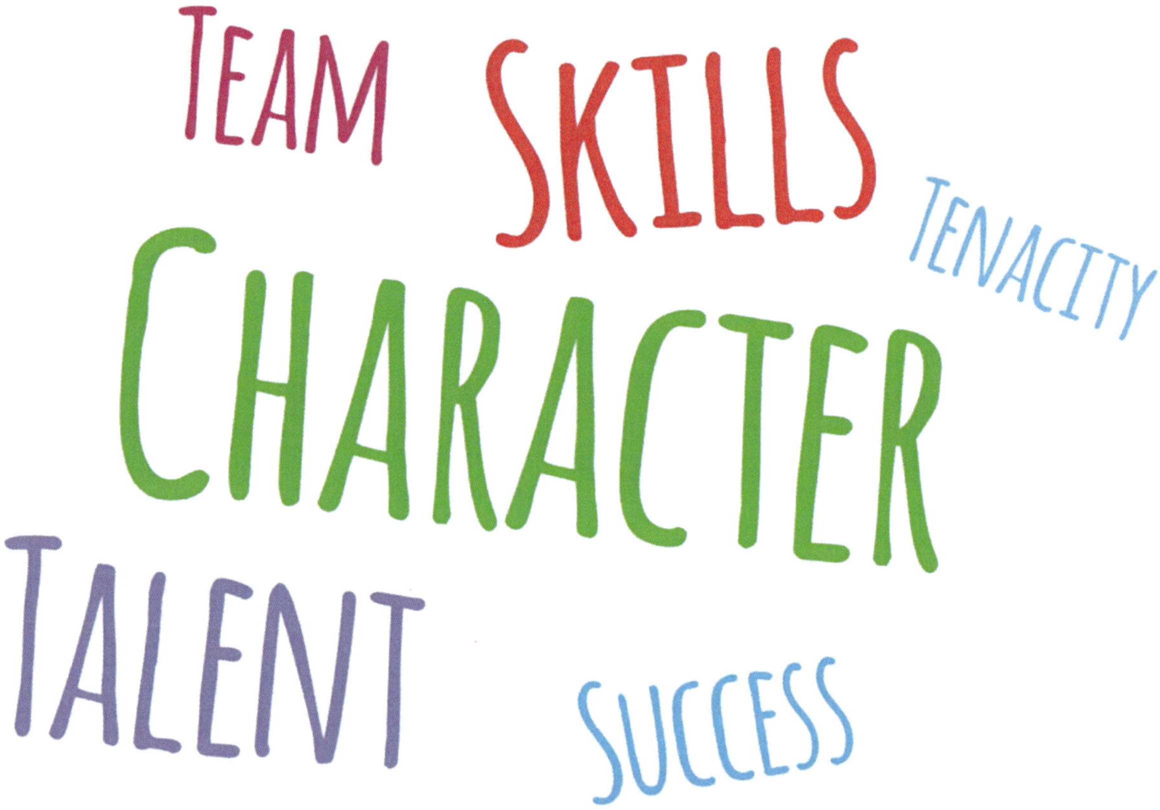

Answer: Only the third statement is True!

The Milkweed, the Monarch, and the Moon book ends with the words "a fresh new start."

What do you think that means? Mark an X in the box next to the 1st or 2nd sentence:

[ ] Forget what happened and do the same thing again.

[ ] Use what you learned to start with a better plan.

Tell about an experience you had that you felt had a beginning, a middle and an end. Explain how something would never be the same again because of what happened in that experience, that made things better in some way than they were at the beginning.

In the 4th section we studied:

Some things we learned about were:

- What kind of character makes a good team member
- How does making a plan help to be successful
- When there is a failure, why the experience still has benefits

Some of the skills you practiced were:

- Listing what is needed for setting a goal
- Making plans
- Being a good team member

On the next few pages you will find puzzles that use the information you learned. Do your best and have fun!

Crossword Puzzle

Solve the clues below and fill in the puzzle. Try to figure out the answers yourself before looking at the answers on the next page.

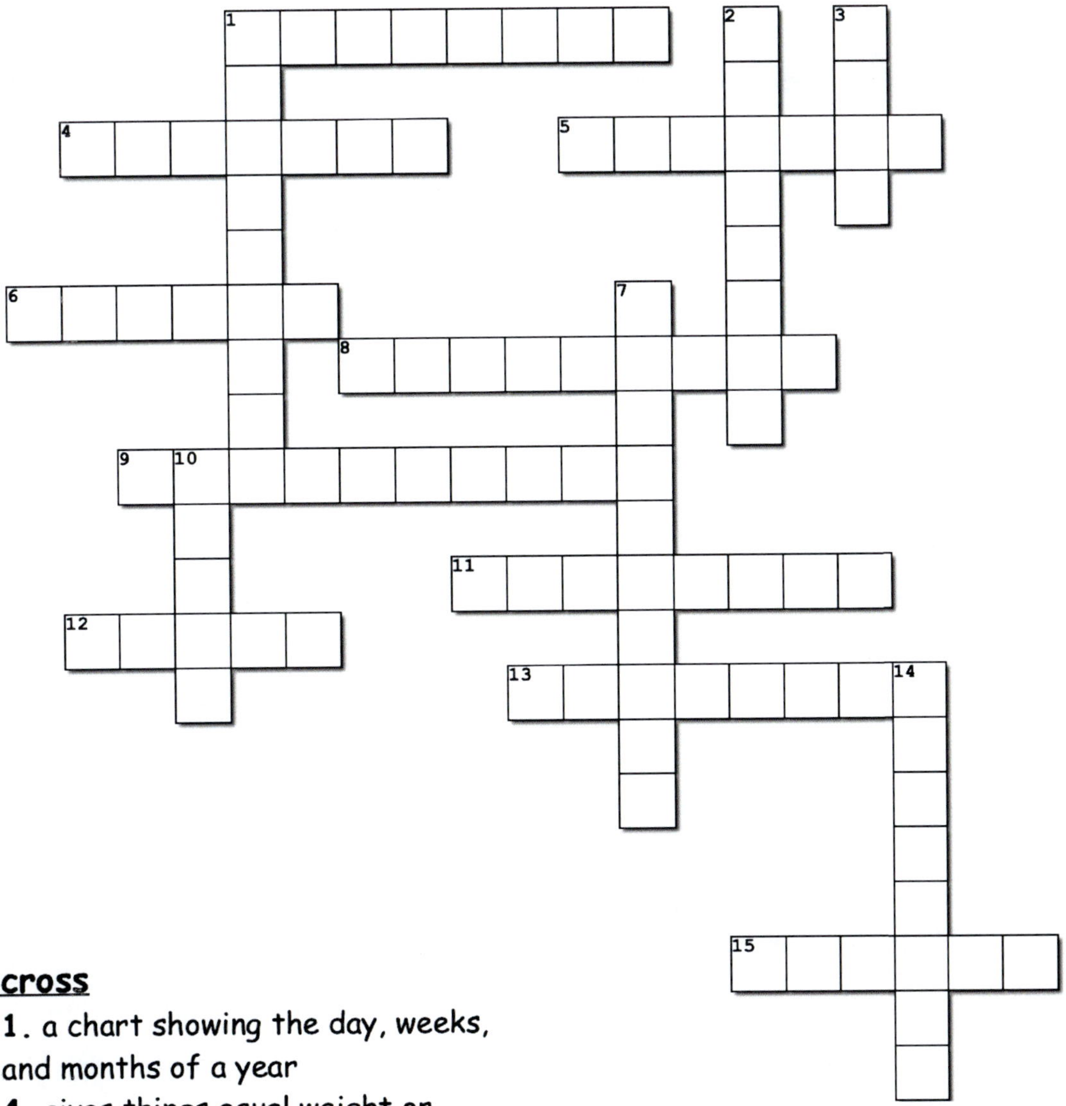

Across

1. a chart showing the day, weeks, and months of a year
4. gives things equal weight or importance
5. a daytime shadow clock
6. a goal you work toward
8. a sand timer
9. the number of hours in a day
11. the repetition of doing something to get better at it
12. when you pay all of your attention to one thing
13. moving forward
15. a unit of time that is equal to sixty seconds

Down

1. a difficulty that is good to overcome
2. a list of jobs for you to complete (3 words)
3. desires that are put into action through a plan
7. a timepiece that wakes you up (2 words)
10. a device you wear on your wrist to tell the time
14. a program of timed events, also to organize a meeting

Answers to the Crossword Puzzle

Across

1. a chart showing the day, weeks, and months of a year (**calendar**)
4. gives things equal weight or importance (**balance**)
5. a daytime shadow clock (**sundial**)
6. a goal you work toward (**target**)
8. a sand timer (**hourglass**)
9. the number of hours in a day (**twentyfour**)
11. the repetition of doing something to get better at it (**practice**)
12. when you pay all of your attention to one thing (**focus**)
13. moving forward (**progress**)
15. a unit of time that is equal to sixty seconds (**minute**)

Down

1. a difficulty that is good to overcome (**challenge**)
2. a list of jobs for you to complete (3 words) (**todolist**)
3. desires that are put into action through a plan (**goal**)
7. a timepiece that wakes you up (2 words) (**alarmclock**)
10. a device you wear on your wrist to tell the time (**watch**)
14. a program of timed events, also to organize a meeting (**schedule**)

Hidden Words

Try to find the words on the list that are related to the subject of the good character and goals. They are hidden horizontally, diagonally, or vertically. The letters may be backwards.

S	U	C	C	E	S	S	C	H	K	M	R	R	T	S	V	Z	F	K	E
K	R	E	X	P	E	C	T	A	T	I	O	N	E	Y	I	Y	I	O	S
P	R	E	D	I	C	T	A	B	L	E	E	L	G	W	N	T	C	O	F
P	W	G	P	Z	Q	I	R	P	D	O	F	G	F	R	A	M	H	C	C
L	S	U	C	C	E	S	S	B	R	E	R	R	A	C	V	R	U	Y	H
E	D	U	T	I	T	A	R	G	S	E	A	T	F	R	N	G	D	Q	A
A	S	Y	N	L	F	C	V	T	V	U	P	E	N	D	U	H	M	E	R
S	J	D	E	O	G	X	E	S	J	I	U	A	V	O	I	O	C	P	A
U	T	I	L	W	V	E	H	X	A	R	C	I	R	R	C	N	C	Z	C
R	A	V	A	Z	M	A	A	C	O	P	E	T	W	E	E	F	K	N	T
E	X	U	T	Q	R	N	C	T	T	F	O	M	O	D	D	S	L	E	E
L	K	N	I	E	E	O	T	Z	Y	W	S	L	I	R	Y	B	E	E	R
T	N	E	V	E	M	U	K	P	N	A	W	F	U	L	Y	J	E	D	S
I	R	F	A	P	X	U	L	V	C	S	N	O	H	N	B	X	I	L	P
N	Y	N	L	Y	X	S	V	N	D	O	I	B	S	K	M	C	H	M	U
W	I	I	S	F	D	W	F	T	C	E	T	U	B	I	R	T	N	O	C
Z	S	T	L	U	S	E	R	I	A	Z	G	V	C	M	E	Z	U	W	R
H	W	K	O	I	C	H	L	D	K	Q	E	C	G	F	I	I	W	R	Z
G	L	X	R	H	M	Q	H	K	P	H	O	E	M	D	Q	F	A	W	S
M	V	X	K	E	X	Q	U	R	P	G	F	E	M	O	T	I	O	N	W

ACCOMPLISH
CHARACTER
CONFIDENCE
CONTRIBUTE
COPE
DESERVE
EMOTION
EVENT
EXPECTATION
GRATITUDE
PLEASURE
PREDICTABLE
PREPARED
RESULT
REWARD
SELFCONTROL
SELFESTEEM
SHARE
SUCCESS
TALENT
VICTORY

You have reached the end of Section 4!

The Beginning of Section 5

Summary Quiz

The following questions can help you remember what you have learned.

1) Which word below checks for progress?
   a) observe
   b) track
   c) spend time
   d) imagine

2) When the moon is full, what is the date on the lunar calendar?
   a) 1st
   b) 15th
   c) 7th
   d) 29th

3) The measure of time that the moon takes to orbit the earth is...

a) a solar year

b) a season

c) a lunar month

d) a lunar week

4) A lunar month has close to how many days?

a) 29

b) 25

c) 35

d) 10

5) Where does the light first appear on the new moon?

a) a crescent on the left

b) a crescent on the right

c) it is a full circle of light

d) it is totally dark

6) A book about the real life of a person is called...

a) history

b) biography

c) science fiction

d) science

7) Why would an author put a message before a story?

a) to give permission to make copies

b) to show honor to someone special

c) to thank anyone who was a valuable team player

d) all the above reasons in a, b, and c

8) When something repeats in an expected way...

a) it is a total surprise

b) it is boring

c) it is a coincidence

d) it is a pattern

9) The name of the thing that can make food from sunrays

a) inanimate object

b) plant

c) animal

d) alien

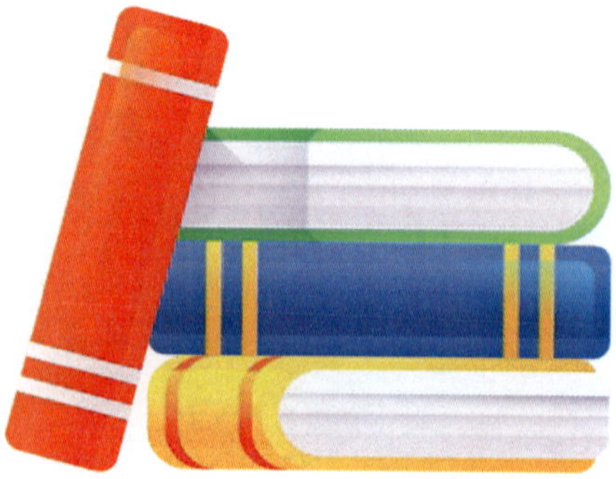

10) Four basic types of team players are...

a) creative thinkers, leaders, organizers, and producers

b) scientists, artists, celebrities, and millionaires

c) magicians, astronauts, politicians, and sports stars

d) family, friends, neighbors, and teachers

11) A Venn diagram...

a) shows sets have nothing in common

b) shows sets have everything in common

c) shows that sometimes sets have somethings in common

d) shows that there are no sets

12) When deciding on what you prefer, it helps to...

a) consider the pros and cons

b) choose whatever is the easiest

c) do whatever your friends are doing

d) just wait for something to happen

13) A biome consists of a natural environment and...

a) climate changes, called seasons

b) features of the earth, called the landscape

c) the native plants and animals, called an ecosystem

d) the way one thing eats another thing, called a food chain

14) The most important things for a human being to survive are...

a) air, water, and the right type of food

b) food, clothes, and games

c) friends, free time, and toys

d) shopping, sports, and adventure

15) A daily planner is used to...

a) count the hours you are asleep

b) show how to do things

c) remind you to try to do things at a set time

d) tell what to wear in your climate

16) The five ways that humans get information are...

a) seeing, hearing, eating, exercising, and sleeping

b) seeing, hearing, touching, tasting, and smelling

c) reading, writing, thinking, sports and sleeping

d) reading, writing, thinking, sleeping, and watching television

17) The five basic biomes are...

a. aquatic, forest, desert, tundra, and grasslands

b. wet, dry, humid, hot and cold

c. spring, summer, fall, autumn and winter

d. windy, rainy, dusty, sunny, and snowy

18) How many hours of sleep are recommended?

a. 7 to 9

b. 10 or more

c. 5 or less

d. there is no recommended amount

19) A quarter of an hour is _____ minutes

a. 15

b. 5

c. 30

d. 60

Good job!

The Contents page can help find what you may want to review.

Correct answers: 1-b, 2-b, 3-c, 4-a, 5-b, 6-b, 7-d, 8-d, 9-b, 10-a, 11-c, 12-a, 13-c, 14-a, 15-c, 16-b, 17-a, 18-a, 19-a

Just for fun, here is a bit of fiction that plays with knowledge about four seasons:

A Week of Fashion

Back in time when messages were mailed on paper, the egg of a Ladybug fell on a page. It was as small as the period at the end of this sentence. A letter writer sent the page to a fashion designer in New York. The letter arrived in the winter.

When the egg hatched, the new Ladybug was named Dot. She wore winter clothes until she traveled south from New York to Atlanta, Georgia. There the climate was warm, so she changed to spring clothes.

She continued on a flight south to the equator where the temperature was hot like summer. She played on the beach in a swimsuit. After that she took one more trip to the hills of Buenos Aires, Argentina, where the climate was cool like in the fall.

You can cut out the paper pal and clothes on the following pages to dress Dot for the four seasons. Also, you may use this idea as a model. For example, you can design clothes for different jobs or for different biomes, depending on your story.

Here are some facts to help make the fiction about Dot believable -

The distance from the earth to the sun is different above or below the equator. The equator is an imaginary ring around planet Earth that marks the middle between the North Pole and the South Pole.

Imaginary circles that climb up and down from the poles like a ladder are latitude lines. (Lines that connect the two poles to measure the surface of the Earth from the east to the west are longitude lines.)

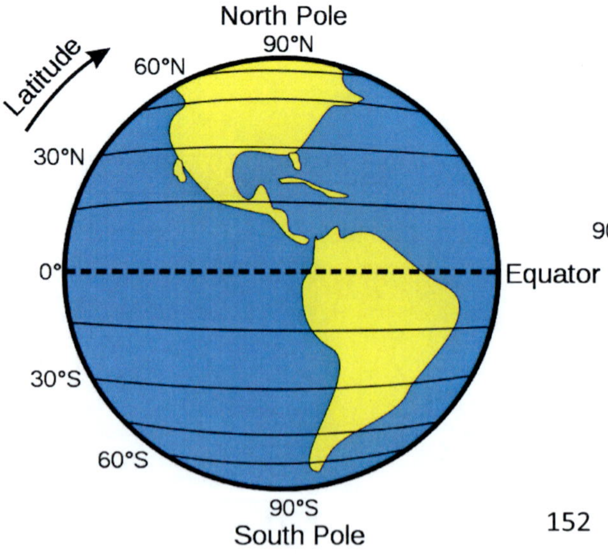

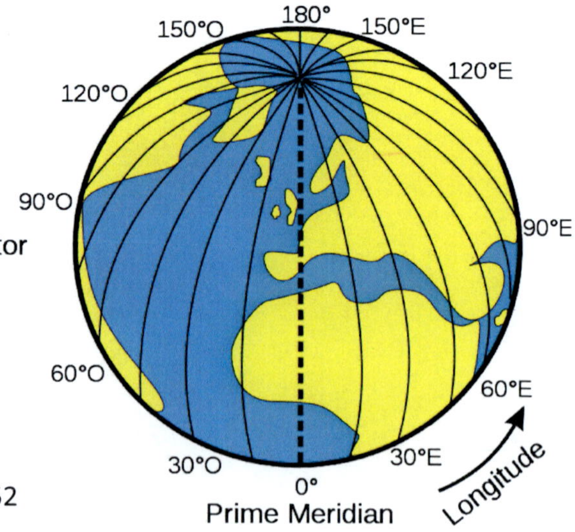

Dot the Paper Pal

Winter

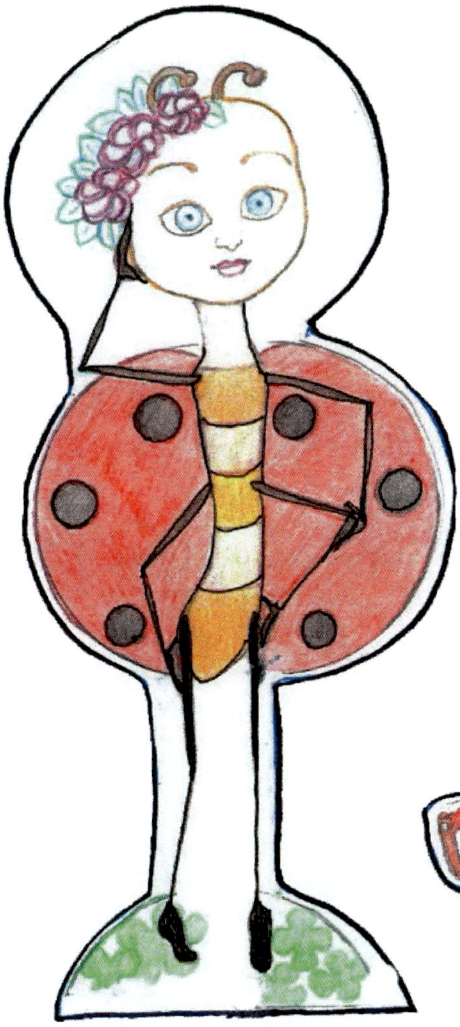

This is a Lady Bug paper pal with clothes for all seasons. With the help from an adult, you can make a photocopy of this page and then cut along the black lines to play with Dot and have her try on different clothes!

Spring

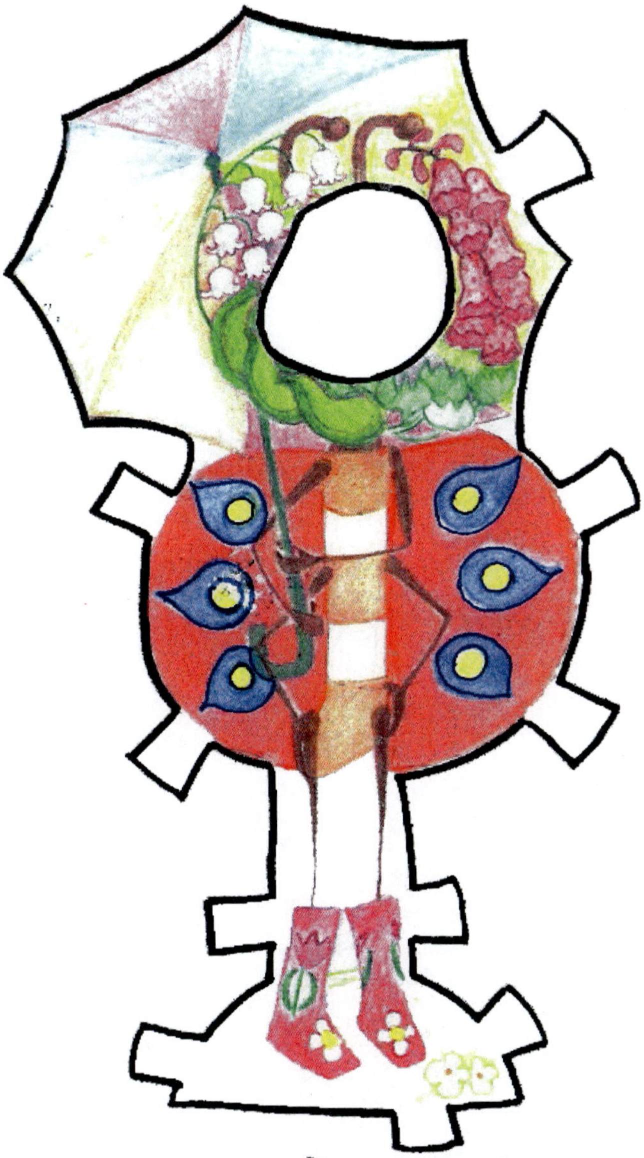

Summer

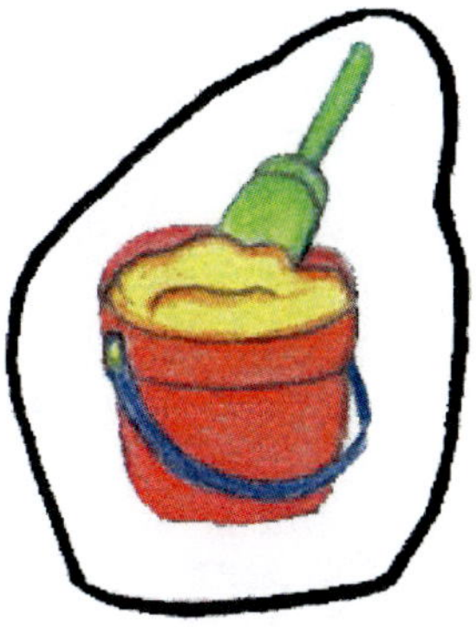

Fall

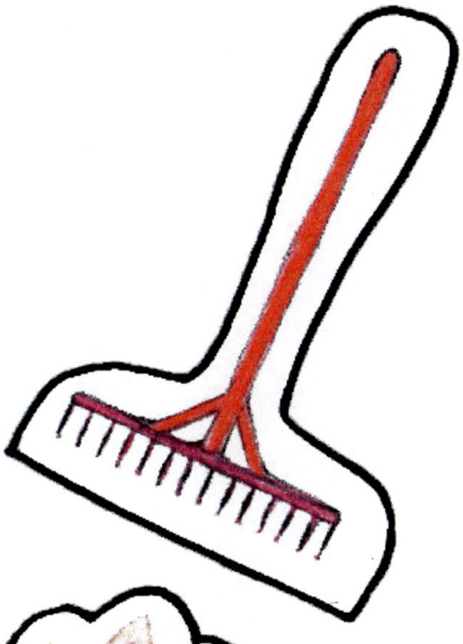

Make Your Own Paper Pal

You can use the models of a paper pal and changes of clothes to illustrate a story of your own!

Write your own story about a character who will change clothes. First, name your character and set a goal. Then, tell what happens as your characters pursues a goal. In the end, tell what your character gained.

Draw, color, and cut out the model to be ready to stage your story.

Pal Template

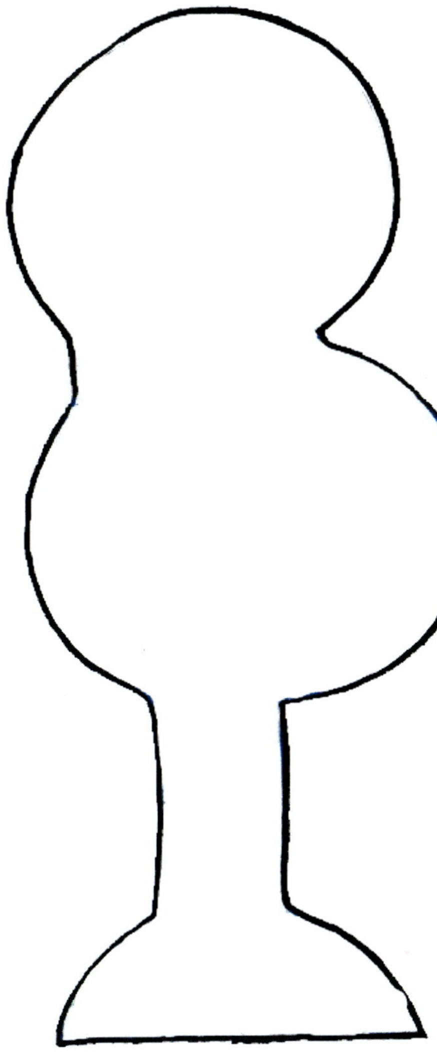

Clothes Template

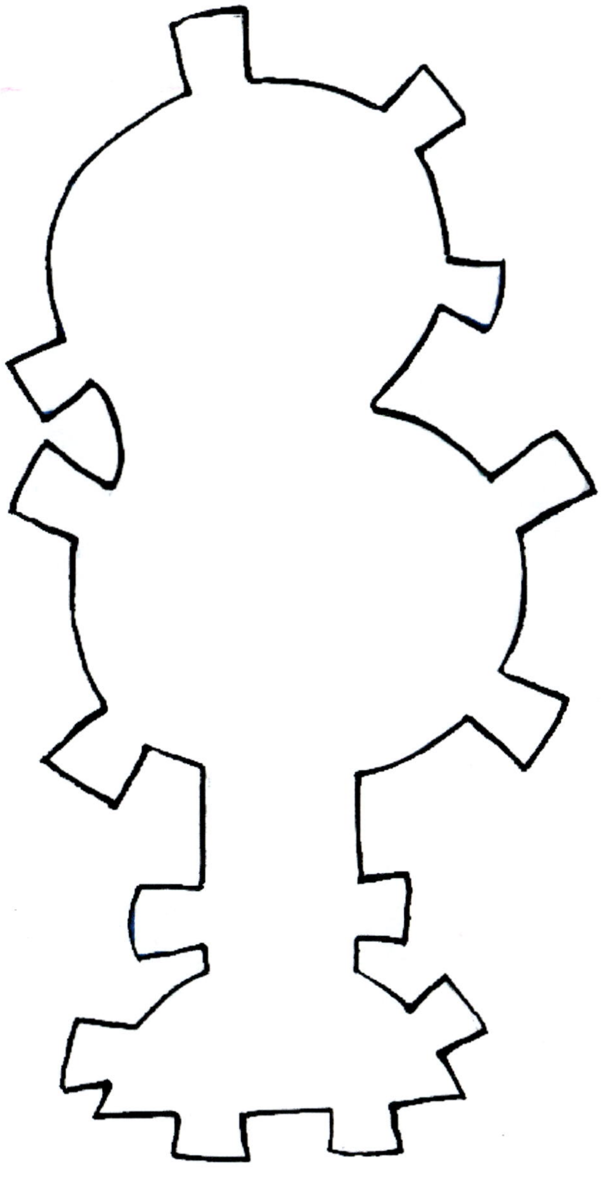

If you enjoyed completing this learning experience...

Congratulations!

Are you more aware of the world around you and things you can do?

I'm excited to think you became more equipped to use time well.

Even if you don't succeed to reach your goal the first time, stay hopeful...

because the pattern of life brings endings and beginnings, there shall be a new moon...

FRESH
NEW
START!

Feedback Form

Your name: ________________________

Put an X to show you are: student [ ] or teacher [ ]

Your grade level: _____

Where do you live (country, state)? ___________________________

Please tell what your favorite section was and why you liked it:

____

____

Also, any ideas for improvements will be appreciated!

____

____

Attach this page to an email and send it to Davida (Galilee) Lippman,

DavidaLippman@gmail.com

Thank you!

Davida and the MMM Team

Made in United States
North Haven, CT
04 November 2026

21731252R00100